Lessons Learned From Maintenance Rule Baseline Inspections

Manuscript Completed: September 1999
Date Published: October 1999

Prepared by
S.-M. Wong, F.X. Talbot, R.M. Latta,
R.P. Correia

Division of Inspections Program Management
Office of Nuclear Reactor Regulation
U.S. Nuclear Regulatory Commission
Washington, DC 20555-0001

ABSTRACT

This report summarizes the lessons learned from 68 maintenance rule (MR) baseline inspections (MRBIs) conducted at plants holding operating licenses in accordance with Title 10, Part 50, of the *Code of Federal Regulations* (CFR) and 3 MRBIs conducted at plants with decommissioning certifications in accordance with 10 CFR 50.82(a)(2). The MRBIs were conducted between July 15, 1996, and July 10, 1998. In general, these MRBIs revealed that licensees implemented the requirements of the maintenance rule, 10 CFR 50.65, by following the guidance in NUMARC 93-01, "Industry Guideline for Monitoring the Effectiveness of Maintenance at Nuclear Power Plants," as endorsed by NRC Regulatory Guide 1.160, "Monitoring the Effectiveness of Maintenance at Nuclear Power Plants." Additionally, licensees effectively determined which structures, systems, and components (SSCs) at each site were within the scope of the MR. The use of expert panels was effective in determining which SSCs were risk significant. The results of the MRBIs also indicated that several expert panels performed other MR activities that exceeded the guidance in NUMARC 93-01. When setting goals or performance measures (criteria) in accordance with 10 CFR 50.65(a)(1) or (a)(2), respectively, most licensees considered risk insights from the probabilistic risk assessments (PRAs). However, early MRBIs revealed that some licensees did not have adequate technical justification for deviating from SSC reliability and availability assumptions in the PRAs when establishing goals or performance criteria and did not adequately assess planned and emergent maintenance activities. Most licensees' individual self-assessments were a MR program implementation strength. Most licensees also established reasonable plans and methods to periodically evaluate the effectiveness of equipment performance monitoring and preventive maintenance, including the balance between reliability and availability.

TABLE OF CONTENTS

APPENDICES

EXECUTIVE SUMMARY

The NRC conducted a maintenance rule baseline inspection (MRBI) at 68 nuclear power plants holding operating licenses and 3 nuclear plants in a decommissioning status. The purpose and objective of the MRBIs were to verify licensees' efforts to (1) establish adequate programs to implement 10 CFR 50.65, the maintenance rule (MR or rule) and (2) determine that the programs were being effectively implemented. Specifically, all sites were inspected in accordance with Inspection Procedure (IP) 62706, "The Maintenance Rule," to determine compliance, strengths, and weaknesses in licensees' implementing programs and the results achieved at each site. This report presents the results of the MRBIs; the major findings for each subject area are summarized below.

Use of Industry Guidelines

All licensees implemented the rule using the guidance in NUMARC 93-01, "Industry Guideline for Monitoring the Effectiveness of Maintenance at Nuclear Plants," Revision 2, dated April 1996, which the NRC endorsed in Regulatory Guide (RG) 1.160, "Monitoring the Effectiveness of Maintenance at Nuclear Power Plants," Revision 2, dated March 1997. Although many licensees took certain exceptions to selected aspects of the guidance in NUMARC 93-01, most licensees presented adequate technical justification for these exceptions. For discussion on exceptions to NUMARC 93-01 for which technical justification was not presented, refer to NRC MR-related inspection reports.

Scoping

Most licensees were thorough in determining which structures, systems, and components (SSCs) were within the scope of the rule at their sites. Licensees at operating plants correctly identified most safety-related SSCs and those non-safety-related SSCs whose failure could prevent safety-related SSCs from fulfilling their safety-related function. However, some licensees did not properly identify a few non-safety-related SSCs as being within the scope of the rule. These non-safety-related SSCs were those relied upon to support the mitigation of accidents or transients or to support the execution of emergency operating procedures. Among the omitted SSCs were some whose failure could cause a reactor scram or actuation of a safety-related system.

Risk Determination Process

The MRBI teams found that each licensee used a well-qualified expert panel to determine the safety (risk) significance of SSCs within the scope of the rule. As described in RG 1.160, the use of an expert panel that considers probabilistic risk assessment (PRA) insights in combination with operating experience, maintenance results, and deterministic evaluations are appropriate and practical methods of determining high safety (risk) significant (HSS) and low safety (risk) significant (LSS) SSCs. The expert panels at many sites considered PRA insights using risk reduction worth (RRW), risk achievement worth (RAW), core damage frequency (CDF) contribution, Fussell-Vesely importance (FVI) and Birnbaum importance (BI) measures. However, some expert panels took exception to the use of the three importance measures (i.e., RAW, RRW, CDF contribution) referenced in NUMARC 93-01 when classifying HSS and LSS SSCs for various reasons. The NRC found each of these cases acceptable since expert panels also used qualitative criteria to determine HSS SSCs that compensated for limitations in the use of PRA insights.

Categorizing Structures, Systems, and Components in Paragraphs (a)(1) or (a)(2)

The process and procedures used by most licensees for categorizing SSCs under paragraph (a)(1) or (a)(2) of the rule were reasonable. However, as indicated during the initial MRBIs, some licensees were reluctant to place SSCs in the paragraph (a)(1) category because they believed that having SSCs in that category would imply that their preventive maintenance programs were ineffective. By the end of the MRBIs, most licensees were appropriately utilizing the paragraph (a)(1) monitoring category to focus their management's attention on SSCs with performance problems that needed corrective actions to improve maintenance effectiveness.

Safety Considerations in Goal Setting and Corrective Actions

Most licensees used risk-informed, performance-based processes and procedures as a primary method of taking safety into consideration when setting goals and taking corrective actions. To implement the maintenance rule, most licensees assigned corrective action responsibility to either the system engineers, the maintenance rule coordinator, or the expert panel. However, some licensees did not establish adequate goals commensurate with safety in all cases. Also, some licensees did not adequately monitor availability for HSS SSCs, which did not allow them to determine the overall performance of the affected equipment.

Industry Operating Experience

Most licensees developed adequate guidance for ensuring that industry-wide operating experience (IOE) was considered when establishing goals. During the last phase of the MRBIs, many plants began sharing IOE information on reliability and availability data for SSCs within the scope of the maintenance rule using the Institute for Nuclear Power Operations (INPO) Equipment Performance and Information Exchange (EPIX) database. Licensees can now use this information to conduct probabilistic risk assessments (PRAs) of equipment performance and to determine if adjustments are needed to reliability and availability goals, performance criteria (measures)[1], and preventive maintenance.

Monitoring and Trending System and Component Performance

Most licensees adequately monitored equipment performance at the system, train, component, or plant level depending upon the safety (risk) significance of each system, train, or component. Similarly, almost

---

[1]Because "goals and monitoring," *per se*, are not required under paragraph (a)(2), NUMARC 93-01 established the term "performance criteria" to describe the measures (e.g., reliability) of performance for SSCs being tracked under (a)(2). However, the NRC uses the term "performance measures," meaning the same thing, in certain MR-related correspondence. For consistency with NRC and industry guidance documents, "performance criteria" will be used in this report.

all licensees appropriately monitored the reliability and availability of HSS
SSCs at the train or component level for those components that have an impact on train functions. Most licensees also monitored the reliability, availability, and/or condition of LSS standby SSCs, where appropriate. Generally, licensees monitored LSS normally operating SSCs using plant level performance criteria (i.e., causes scrams, safety system actuations, or unplanned capacity loss factors).

Monitoring and Trending the Condition of Structures

Prior to the first MRBIs, the NRC and NEI agreed that additional structural monitoring guidance was needed to meet the intent of the rule. The NRC revised RG 1.160, Revision 2, by adding Regulatory Position 1.5, "Monitoring Structures." This guidance provided additional condition monitoring methods for structures and established a technical basis for determining when a structure should be re-categorized into the paragraph (a)(1) monitoring category with more focus on predictive monitoring techniques. After the NRC issued RG 1.160, Revision 2, licensees updated their structural monitoring programs to include monitoring and goals that were predictive in nature, providing a better means of identifying degradation before failures occurred.

Periodic Evaluations

Most licensees completed the periodic evaluations required by the rule within the prescribed time frame. However, a few licensees were late in completing the initial periodic evaluation. Most licensees used IOE experience to evaluate the reliability and availability of their SSCs common across the industry and made adjustments to preventive maintenance (PM) activities. It was also noted during the MRBIs that many licensees used the periodic evaluations to determine the overall efficacy of their MR implementation programs and made adjustments to PM activities, where necessary, to improve overall equipment performance.

Balancing Reliability and Unavailability

Most licensees established appropriate methods for balancing system and component reliability versus unavailability. Most licensees used PRA to determine the appropriate balance between system and component reliability and unavailability for HSS SSCs. In particular, if HSS SSC performance was maintained within performance criteria limits based on risk analysis assumptions and operating experience, then licensees achieved a balance between reliability and availability. A few licensees used PRA conditional probability methods to establish bounding limits for reliability and availability. If properly applied, this method can also be used to determine whether a balance is being achieved between reliability and availability for HSS SSCs.

Safety (Risk) Assessments Before Performing Maintenance

Licensees developed various qualitative and quantitative methods to determine the risk associated with performing maintenance while the plant was operating and during shutdown conditions. These methods were: (1) using matrices of pre-analyzed combinations of two SSCs, which were used to evaluate the risk associated with performing maintenance on them at the same time, (2) using PRA specialists to provide risk insights in the scheduling and planning process to pre-analyze unknown risk for combinations of three or more SSCs, (3) using planning and scheduling procedures that maintain control of rolling weekly schedules with pre-analyzed maintenance configurations, and (4) using sophisticated near-real-time risk monitors to calculate the risk changes associated with planned maintenance activities. In addition, most licensees established adequate controls for emergent work activities where contingency actions could be taken to change a planned schedule and prioritize returning equipment to service.

The MRBI teams found 26 pre-maintenance safety/risk assessment programs that were well established with few or no weaknesses associated with these activities. The MRBI teams found 35 assessment programs with some weaknesses but found no instances where assessments were not performed before conducting maintenance on SSCs. The MRBI teams also found seven licensees with weak assessment programs where safety assessments were not performed before conducting maintenance on SSCs. These findings and changes in the amount and frequency of maintenance performed at power were some of the reasons the NRC staff proposed to the Commission that pre-maintenance assessments should be changed from a recommendation to a requirement.

Decommissioning Status Plants

The NRC completed three MRBIs at plants in a decommissioning status in accordance with 10 CFR 50.82. The results of these MRBIs indicated that two of three plants had acceptable maintenance rule monitoring programs on SSCs needed to store, control, and maintain the spent fuel in a safe condition.

Conclusions

The MRBI teams concluded that the requirements of 10 CFR 50.65 can be met using NUMARC 93-01 as endorsed by RG 1.160, Revision 2. The teams also concluded that using PRA insights in conjunction with an expert panel, provided acceptable methods for taking risk into consideration when implementing the rule. IP 62706 was found to be adequate to evaluate the acceptability of licensees' maintenance rule programs and their implementation.

The MRBI teams further concluded that the risk-informed, performance-based approach to implementing the maintenance rule was practical; however, the NRC staff also concluded that future NRC inspections should focus on performance issues using risk insight and, where necessary, programmatic aspects of the rule, if findings indicate that causes for performance problems are rooted in the MR program.

ACKNOWLEDGMENTS

The authors want to acknowledge all regional team leaders, Office of Nuclear Reactor Regulation (NRR) staff support members, specialist inspectors with expertise in probabilistic risk assessments (PRAs), and regional inspectors who contributed a substantial amount of effort to verify licensees' implementa-tion of the result-oriented, risk-informed, and performance-based maintenance rule. The authors also acknowledge Paula Hollingsworth, Rayleona Sanders, and Stephen Alexander, who edited this report.

Authors

Richard P. Correia, Robert M. Latta, Francis X. Talbot, See-Meng Wong

NRC Team Leaders, NRR Staff Support Members, and Team Members/Inspectors for 68 MRBIs at Operating Nuclear Power Plants and 3 MRBIs at Plants in a Decommissioning Status

Note: Some names appear more than once since the individuals served in more than one role while on inspection teams (i.e., team leader, staff support member, specialist inspector with PRA expertise, and/or team inspector).

Team Leaders	NRR Staff Support Members
Paul H. Bissett	Donnie J. Ashley
Larry E. Briggs	Peter A. Balmain
Richard P. Correia	Thomas A. Bergman
John G. Caruso	Michael T. Bugg
Billy R. Crowley	Ronald K. Frahm, Jr.
Andrew Dunlop	Edward J. Ford
Lawrence E. Ellershaw	Kenneth C. Heck
Martin J. Farber	Robert M. Latta
Donald J. Florek	Charles D. Petrone
Paul C. Gage	Wayne E. Scott
Ronald D. Gibbs	Francis X. Talbot
William Holland	Donald R. Taylor
Claude E. Johnson	Steve G. Tingen
Thomas J. Kenny	J. D. Wilcox, Jr.
Charles J. Paulk	Peter R. Wilson
Julian H. Williams	See-Meng Wong
J. D. Wilcox, Jr.	
John E. Whittemore	

ACKNOWLEDGMENTS

Team Members/Inspectors

Russell J. Arrighi
Donnie J. Ashley
Hansraj G. Asher
Ramon V. Azua
Stephen T. Barr
William C. Bearden
Thomas A. Bergman
Rudolph H. Bernhart
Danny E. Billings
Paul H. Bissett
Larry E. Briggs
Carey E. Brown
Charles H. Brown
Michael T. Bugg
Sonia D. Burgess
P. Byron
James A. Canady
Robert K. Caldwell
Joseph E. Carrasco
John G. Caruso
Dr. Jin W. Chung
Clifford K. Clark
Douglas H. Coe
James L. Coley
Laura L. Collins
William A. Cook
Kevin A. Coyne
Joseph M. D'Antonio
Steven Dennis
Glenn T. Dentel
Binoy B. Desai
Stephen C. Dinsmore
Nickolas Economos
Thomas R. Farnholz
Ricardo A. Fernandez
Todd H. Fish
Donald J. Florek
Edward J. Ford
Michael X. Franovich
Thomas R. Fredette
Timothy J. Frye

Ronald D. Gibbs
Mark A. Giles
Edwin H. Gray
Katherine S. Green-Bates
Eugene F. Guthrie
Samuel L. Hansell, Jr.
James K. Heller
Claude E. Johnson
Donald E. Jones
Steven R. Jones
William B. Jones
William P. Kleinsorge
Kenneth S. Kolaczyk
John G. Kramer
Ronald A. Langstaff
David R. Lanyi
Henry K. Lathrop
Robert M. Latta
Samuel S. Lee
Alfred Lohmeier
Raymond K. Lorson
John S. Ma
William M. McNeil
James F. Melfi
Rogelio Mendez
Merle N. Miller
James H. Moorman III
John H. Neisler
Randall A. Musser
Kathleen F. O'Donohue
Gerald F. O'Dwyer
Bradley J. Olsen
Clyde C. Osterholtz
Michael E. Parker
Charles J. Paulk, Jr.
David L. Pelton
Charles D. Petrone
Gregory A. Pick
Stephen M. Pindale
Robert J. Prato
George D. Replogle
Jose Rogerio Reyes
Daniel W. Rich

John E. Richmond
Darrel J. Roberts
Walter G. Rogers
Steve P. Sanchez
Wayne E. Scott
Darrel L. Schrum
Scott C. Schwind
Jeffrey L. Shackelford
John T. Shedlosky
Carl E. Sisco
Richard A. Skokowski
Christopher E. Skinner
Scott E. Sparks
Joelle L. Starefos
Stanle Stasek
Francis X. Talbot
Donald R. Taylor
Tirupataiah Tella
Ross D. Telson
James M. Trapp
Glen Walton
Kathy D. Weaver
Gregory E. Werner
Geoffrey Wertz
Herbert L. Whitener
John E. Whittemore
J. D. Wilcox
Julian H. Williams
Peter R. Wilson
See-Meng Wong

Consultants — Specialist Inspectors With PRA Expertise

Michael Calley
Richard Deem
Adele DiBiasio
Steve Eide

Anthony Frescoe
Willie Galyean
Dana Kelly
Gerardo Martinex
Curtis Smith
Richard Travis
Wei He

Observers From Foreign Regulatory Authorities

P.C. Ambros, National Commission of Nuclear Energy, Brazil
Yves Balloffet, Directorate for the Safety of Nuclear Installations, France
Julio Crespo Bravo, Council of Nuclear Safety, Spain
Luis A. Germez Martin, Consultant for the Council of Nuclear Safety, Spain
Angel L. Coello Ortega, Council of Nuclear Safety, Spain

Headquarters and Regional Maintenance Rule Program Supervisors

Goutam Bagchi
Harold O. Christensen
Richard J. Conte
Richard P. Correia
Paul E. Fredrickson
James A. Gavula
John Jacobson
WIlliam Holland
Wayne Kropp
Glenn W. Meyer
Dale A. Powers
Mark Rubin
Dave Verrelli

ABBREVIATIONS

AEOD	Office for the Analysis and Evaluation of Operational Data
ACRS	Advisory Committee on Reactor Safeguards
ANS	American Nuclear Society
ATWS	anticipated transient without scram
BI	Birnbaum importance
BWR	boiling-water reactor
CDF	core damage frequency
CFR	*Code of Federal Regulations*
CP	conditional probability
CR	condition report
DG	draft regulatory guide
DMD	number of demands
DRS	Division of Reactor Safety
DSI	direction-setting issue
E/C	erosion/corrosion
ECCS	emergency core cooling system
EDG	emergency diesel generator
EGM	enforcement guidance memorandum
EMEB	Materials and Chemical Engineering Branch
EOOS	equipment out of service
EOP	emergency operating procedure
EPIX	Equipment Performance and Information Exchange
EPRI	Electric Power Research Institute
ESFAS	engineered safety features actuation system
FF	functional failure
FVI	Fussell-Vesely importance
FR	*Federal Register*
FSAR	final safety analysis report
GE	General Electric
GSW	general service water
HEP	human error probability
HQMB	Quality Assurance, Vendor Inspection and Maintenance Branch
HSS	high safety (risk) significant

ABBREVIATIONS

I&C	instrumentation and control
IN	information notice
INPO	Institute for Nuclear Power Operations
IOE	industry-wide operating experience
IP	inspection procedure
IPE	individual plant examination
IPEEE	individual plant examination of external events
IQMB	Quality Assurance, Vendor Inspection, Maintenance and Allegations Branch
LERF	large early release frequency
LSS	low safety (risk) significant
MEF	mean expected number of failures
MOV	motorized operated valve
MPFF	maintenance preventable functional failure
MPRFF	maintenance preventable redundancy functional failure
MR	maintenance rule
MRBI	maintenance rule baseline inspection
MRIC	Maintenance Rule Inspection Clearinghouse
MTI	maintenance team inspection
NEI	Nuclear Energy Institute
NPRDS	nuclear plant reliability data system
NRC	Nuclear Regulatory Commission
NRR	Office of Nuclear Reactor Regulation
NSAL	nuclear service advisory letter
NSSS	nuclear steam supply system
NSTB	nuclear service technical bulletin
NUMARC	Nuclear Management and Resources Council
OE	Office of Enforcement
OGC	Office of the General Counsel
OOS	out-of-service
PGEB	Generic Issues and Environmental Projects Branch
PRA	probabilistic risk assessment
PSA	probabilistic safety assessment
QA	quality assurance
Q&As	questions and answers
RAW	risk achievement worth
RG	regulatory guide
RHR	residual heat removal

ABBREVIATIONS

RPS	reactor protection system
RRW	risk reduction worth
SECY	Secretary of the Commission
SER	significant event report
SES	senior executive service
SFPCC	spent fuel pool cooling and cleanup
SIL	service information letter
SOCs	statements of consideration
SOER	significant operating experience report
SPSB	Probabilistic Safety Assessment Branch
SRA	senior reactor analyst
SRM	staff requirements memorandum
SSM	staff support member
SSC	structure, system, and component
TDFWP	turbine-driven feedwater pump
TIL	technical information letter
TR	topical report
TSs	technical specifications
URL	uniform resource locator
V&V	verification and validation

1 INTRODUCTION—MAINTENANCE RULE BASELINE INSPECTIONS

On July 10, 1991, the U.S. Nuclear Regulatory Commission (NRC) published the maintenance rule (MR or the rule) in the *Federal Register* (56 FR 31324) as Title 10, Section 50.65, "Requirements for monitoring the effectiveness of maintenance at nuclear power plants," of the *Code of Federal Regulations* (10 CFR 50.65). The FR notice stated that this rule must be implemented by each licensee no later than July 10, 1996.

The NRC staff determined that it should conduct maintenance rule baseline inspections (MRBIs) at all operating plants within a 2-year period following the effective date of the rule. The four NRC regional offices effectively and efficiently planned, scheduled, and completed this task within the prescribed time frame in collaboration with the Office of Nuclear Reactor Regulation, Quality Assurance, Vendor Inspection and Maintenance Branch.

Although the rule was originally promulgated as a results-oriented, performance-based regulation, the primary intent of the MRBIs was to determine the adequacy of the licensees' MR programs and processes to implement the program. This program-oriented approach ensured consistent application of the rule, verified that licensees had effective MR performance monitoring programs, and confirmed that licensees adjusted their MR activities and programs when performance indicated the need for a change. This program-oriented approach also allowed the NRC and the industry to gain experience with risk-informed, performance-based regulation, which can be used to develop other risk-informed, performance-based regulatory approaches to such programs as technical specifications, in-service testing, in-service inspection, and graded quality assurance.

Accordingly, this report discusses the results and lessons learned insight gained from 68 MRBIs conducted by the NRC staff at operating commercial nuclear power reactor sites and the 3 MRBIs that were performed at commercial nuclear power reactor sites in a decommissioning status pursuant to 10 CFR 50.82.

2 LESSONS LEARNED FROM MAINTENANCE RULE BASELINE INSPECTIONS

Each maintenance rule baseline inspection (MRBI) was unique, involving NRC review of each licensee's site-specific programs and processes. Every licensee implemented the rule using the guidance in NUMARC 93-01, "Industry Guideline for Monitoring the Effectiveness of Maintenance at Nuclear Power Plants," as endorsed by NRC Regulatory Guide (RG) 1.160, "Monitoring the Effectiveness of Maintenance at Nuclear Power Plants"; however, each licensee developed different site-specific programs and processes. This individualized approach presented a challenge to NRC inspectors because they had to evaluate each licensee's program on its own merits and determine whether it satisfied the requirements of the rule. Some licensees took exception to selected aspects of the guidance in NUMARC 93-01, but, in most cases, they presented an adequate technical basis for their site-specific exceptions. The unique nature of each program stemmed first from variations in the numbers and types of structures, systems, and components (SSCs) to consider for inclusion within the scope of the rule; even among plants of similar design; therefore, these plants had variations in the number and type of SSCs ultimately included within the scope of the rule. Among these, there were variations in the SSCs that licensees classified as high-safety-significant (HSS) SSCs or low-safety-significant (LSS) SSCs, as well as substantial variations in categorizing SSCs to be monitored in accordance with paragraph 50.65(a)(1) or to be tracked in accordance with paragraph 50.65(a)(2) as long as effective preventive maintenance could be demonstrated. Finally, there were some differences in the goals set for monitoring SSCs under paragraph (a)(1) and the performance criteria (measures)[1] developed for tracking SSCs performance under paragraph (a)(2). However, this variation was anticipated and is consistent with performance-based regulations and the NRC's intention to give licensees flexibility in implementing the rule.

The NRC staff completed the MRBIs within the prescribed 2 years by dedicating sufficient resources to the inspection effort. Prior to the MRBIs, the Quality Assurance, Vendor Inspection and Maintenance Branch (HQMB) gave extensive training to inspectors on the requirements of the MR and the guidance contained in RG 1.160, NUMARC 93-01, and Inspection Procedures (IP) 62706, "The Maintenance Rule," and IP 62707, "Maintenance Observations." This effort also ensured effective communication among the NRC regional offices, the Office of Nuclear Reactor Regulation (NRR), the Office of Enforcement (OE) and the Office of the General Counsel (OGC). The findings and lessons learned from the MRBIs follow:

Findings

In general, the NRC staff found that licensees adequately implemented the requirements of the rule. However, despite the 5-year implementation period, the MRBIs revealed that several licensees waited until the year before the effective date of the rule to aggressively pursue implementation. Also, some

---

[1]Since "monitoring" *per se* is not required under 50.65(a)(2), NUMARC 93-01 established the term "performance criteria" to describe measures (e.g., reliability) of performance for SSCs being tracked under paragraph (a)(2) of the rule. The NRC uses the term "performance measures" to mean the same thing in certain MR-related correspondence. However, "criteria" will be used in this report for consistency with NRC and industry guidance documents.

licensees did not fully implement their programs until a few months before the inspection effort began. For many of those licensees, the baseline inspections identified weak programs or weak implementation of those programs or both.

The next two paragraphs describe the most significant and most often identified findings of the MRBIs:

(1)   Many licensees did not have an adequate technical basis to demonstrate that goals and performance criteria were established commensurate with safety, or they did not have an adequate technical basis for the values chosen to show that they were statistically linked to the assumptions in the plant risk analyses. For example, some licensees allowed each SSC a standard number of maintenance preventable functional failures (MPFFs) each refueling cycle as a reliability performance criterion without considering the number of demands on the SSC. In some cases, when compared to the number of demands, the number of MPFFs allowed would indicate much lower reliability than the licensee assumed in its risk analyses without adequately justifying the difference in values. The NRC staff took the position that licensees must have a sound technical basis for the reliability and availability values chosen as (a)(1) goals or (a)(2) performance criteria.

(2)   Many licensees did not develop both reliability and availability goals or performance criteria for high-safety-significant (HSS) SSCs. As part of the periodic evaluations required by 10 CFR 50.65(a)(3), licensees must balance reliability and availability. This balance cannot be attained unless the licensee monitors both of these parameters.

The problems described next were found at many sites and were considered significant, although they did not occur as often as the two problems noted above:

(3)   In a few instances, licensees did not include safety-related SSCs or safety-related SSC functions within the scope of the rule in accordance with 50.65(b)(1). For example, one licensee did not include the fuel assemblies while another licensee did not include a high pressure coolant injection pump turbine function. In some cases, licensees did not include non-safety-related SSCs in accordance with 50.65(b)(2). For example, one licensee did not include the cooling tower system within the scope even though a failure of the system had resulted in a scram on one occasion and a transient in another. In other instances, licensees did not include communications or emergency lighting systems in scope, even though industry operating experience demonstrated their importance in supporting plant personnel in mitigating accidents and transients and completing operator actions required by emergency operating procedures. For additional details on scoping issues, see Section 2.3 of this report.

(4)   The MRBI teams also observed that some licensees appeared reluctant to identify failures as MPFFs. In many cases, this reluctance was attributed to the licensees' belief that identifying MPFFs and the possible resultant SSCs in the (a)(1) monitoring category indicated an ineffective maintenance program. From a regulatory standpoint, the occurrence of a single MPFF is not necessarily a violation and is not a violation of the maintenance rule. Rather, an MPFF usually indicates a potential problem and what is more important is that licensees determine the cause and take effective corrective actions to preclude recurrence. Ironically, this reluctance to identify

MPFFs resulted in some maintenance rule violations for not properly monitoring the performance of or demonstrating effective preventive maintenance of SSCs within the scope of the rule.

(5)   Just prior to the first set of MRBIs, it became apparent that there was inadequate NRC and industry guidance to monitor the condition of structures to comply with the requirements of the rule. Accordingly, the inspectors did not have sufficient information to determine whether licensees' structural monitoring programs satisfied the rule requirements. Specifically, the MRBI teams identified 15 sites with structural monitoring program concerns because of inadequate guidance; therefore, the teams opened inspector followup items to reevaluate structural monitoring programs after proper guidance was developed. The NRC staff provided MR specific structural monitoring guidance in RG 1.160, Revision 2, dated March 1997. Once the NRC staff issued RG 1.160, Revision 2, most licensees developed and implemented structural monitoring programs that followed this guidance and met the requirements of the rule.

The NEI (formerly NUMARC) also recognized the need for additional guidance on structural monitoring before the effective date of the rule. NUMARC 93-01, Revision 2, Section 10.2.3, "Monitoring the Condition of Structures," added some guidance. The NEI also developed draft comprehensive guidance in NEI 96-03, "Guideline for Monitoring the Condition of Structures at Nuclear Power Plants" (NEI, 1996). NEI 96-03 intended to provide structural monitoring guidance for all regulatory applications, not just for the maintenance rule. The NRC staff could not complete its review of NEI 96-03 because it believed that NEI 96-03 did not contain sufficient information for all regulatory applications addressed in the guideline.

(6)   Finally, most licensees performed adequate safety assessments before performing maintenance as recommended by 10 CFR 50.65(a)(3). The MRBI teams found 26 pre-maintenance safety/risk assessment programs that were well established with few or no weaknesses associated with these activities. The MRBI teams found 35 assessment programs with some weaknesses, but no instances of an assessment not being performed before performing maintenance on SSCs. However, the MRBI teams found 7 cases in which licensees with weak safety assessment programs had not performed safety assessments before performing maintenance on SSCs. Because this provision of the rule is not an explicit requirement (i.e., the rule states that licenses *should* perform the safety assessments), the NRC staff could not take enforcement action.

Conclusions

During early MRBIs, weaknesses concerning acceptable methods of implementing the requirements of the maintenance rule were identified. Some licensees' late implementation of the MR led to program weaknesses. The MRBI teams identified 10 CFR 50.65(b) compliance issues at several sites. Many sites did not demonstrate that goals or performance criteria or both were commensurate with safety and did not have an adequate technical basis for the values chosen that were statistically linked to the assumptions in their risk analyses. 10 CFR 50.65(a)(3) directs licensees to balance reliability and availability, but some licensees did not develop goals or performance criteria in terms of both reliability and availability for HSS SSCs. This balance cannot be attained unless reliability and availability are monitored. Some licensees appeared reluctant to identify failures as MPFFs. In some cases, this resulted in violations for not adequately monitoring performance. Early MRBIs also identified a number of sites with inadequate structural monitoring programs because of insufficient structural monitoring guidance; therefore, RG 1.160 was revised to address this issue. Many licensees performed adequate safety assessments before performing maintenance on equipment; however, in seven cases, licensees did not perform safety assessments before conducting maintenance. In addition, many safety assessment programs had implementation weaknesses.

Recommendations

The NRC adopted the proposed revisions to 10 CFR 50.65 to delete the last sentence of paragraph 50.65(a)(3) and to require licensees to perform safety assessments before performing maintenance in accordance with the new paragraph 50.65(a)(4). The NRC staff should work with the NEI, the industry, and stakeholders[2] to discuss and develop appropriate revisions to NUMARC 93-01 and RG 1.160.

2.1 Use of Industry Guideline NUMARC 93-01 as Endorsed by RG 1.160

When implementing the maintenance rule using NUMARC 93-01, a licensee first uses the criteria in 10 CFR 50.65(b) to determine which SSCs are within the scope of the rule. The licensee then categorizes all of the SSCs within the scope of the rule as either high safety significant (HSS) or low safety significant (LSS). It is important to emphasize that licensees determine the scope of SSCs under the rule using the deterministic criteria in paragraph (b). For purposes of treatment of SSCs under the scope of the rule, safety (risk) significance is considered only after the scope has been established.

The HSS SSCs monitored under 10 CFR 50.65(a)(1) or evaluated under 10 CFR 50.65(a)(2) should be monitored or tracked at the system, train, or component level. At a minimum, this monitoring should include SSC reliability, availability, and/or condition (i.e., parameter) monitoring. SSCs that are LSS and in standby should also be monitored at the system, train, or component level. Licensees can monitor the reliability, availability, or condition of LSS standby SSCs but, at a minimum, this monitoring should include reliability. SSCs that are LSS and normally operating may be monitored at the plant level. This typically includes monitoring unplanned scrams, safety system actuations, and unplanned capability loss factors.

Those SSCs for which the licensee has demonstrated effective preventive maintenance are tracked in accordance with paragraph (a)(2), as allowed by the rule. For SSCs tracked in accordance with paragraph (a)(2), licensees must establish appropriate performance measures[3]. The only exceptions to not monitoring performance is SSCs that are considered to be inherently reliable or can be run to failure. Performance measures must have an adequate technical basis for the values chosen so that SSC

---

[2] The meaning of the term "stakeholder" as used in this document is "parties with a vested interest in effective and efficient implementation of the MR (e.g., the industry and the public)."

[3] NUMARC 93-01 uses goals and performance criteria with a specific meaning. Goals are used for SSCs monitored under paragraph (a)(1), while performance criteria are used to demonstrate effective preventive maintenance for SSCs under paragraph (a)(2).

performance can demonstrate effective preventive maintenance. By contrast, those SSCs for which the licensee cannot demonstrate effective preventive maintenance are more effectively monitored in accordance with paragraph (a)(1). For SSCs monitored in accordance with paragraph (a)(1), licensees must establish appropriate goals. Accordingly, the goals must be commensurate with safety. In accordance with paragraph (a)(1) of the rule and the guidance in NUMARC 93-01, most licensees monitor SSCs with performance problems until it is demonstrated that performance improvements have been implemented and the SSCs can be returned to the Paragraph (a)(2) good health status.

Provided that an SSC meets its performance measures or does not experience a repetitive MPFF[4], the preventive maintenance for that SSC is considered to be effective, and tracking can continue in accordance with paragraph (a)(2). When the SSC does not meet a performance criterion or experiences a repetitive MPFF, the licensee must determine the cause and whether the SSC should be monitored in accordance with paragraph (a)(1). In addition to monitoring reliability, availability, or both, SSCs monitored under paragraph (a)(1) are expected to have goals that specifically address the cause of the problem that resulted in the SSC being monitored in accordance with paragraph (a)(1).

Licensees are expected to monitor the performance or condition of SSCs within the scope of the rule against the goals and performance measures on an ongoing basis. As required by paragraph (a)(3), licensees must evaluate the effectiveness of maintenance at least once per refueling cycle, not to exceed 24 months between evaluations.

Revision 2 to RG 1.160

During the MRBIs, it became apparent that the guidance documents (i.e., RG 1.160 and NUMARC 93-01) would require revision to reflect lessons learned. The need for revisions to the guidance documents (iterative process) arises from the general requirements of the rule versus specific guidance in the guidance documents and the flexibility given licensees to implement the requirements of the MR.

Findings

The industry conducted a verification and validation (V&V) program, and the NRC staff conducted a pilot program to assess, in part, the adequacy of the inspection procedures and implementation guidance documents before the effective date of the maintenance rule. Following those programs, NEI issued Revision 2 to NUMARC 93-01 (April 1996). The NRC staff subsequently issued draft regulatory guide (DG) 1051 (the proposed RG 1.160, Revision 2) in August 1996, with a public comment period that ended on November 15, 1996. However, the NRC staff did not issue RG 1.160, Revision 2, until March 1997 because of its desire to incorporate lessons learned from the initial MRBIs.

Revision 2 to RG 1.160 endorses and clarifies the guidance presented in NUMARC 93-01, Revision 2. It also incorporates clarifications that resulted from public comments regarding DG-1051, experience with

---

[4]The MPFF is a construct of NUMARC 93-01, defined as a failure that results in a loss of the function that caused the SSC to be within the scope of the rule (e.g., the failure resulted in a scram) that could have been prevented through more effective maintenance prior to the failure.

the baseline inspections, and two public meetings held on October 15, 1996, and January 9, 1997. The paragraphs that follow described the most significant clarifications incorporated in Revision 2 to RG 1.160.

- <u>Changes to the Rule</u>: The maintenance rule was amended twice since RG 1.160, Revision 1 was issued.[5] On August 28, 1996, the rule was amended to specifically address the SSCs within the scope of the rule for plants in a decommissioning status. On December 11, 1996, as part of the final rulemaking for 10 CFR Part 50, Appendix S, "Earthquake Engineering Criteria for Nuclear Power Plants," the NRC changed the definition of "safety-related" SSCs in paragraph (b)(1) of the maintenance rule to make it consistent with the use of that term in other regulations. However, neither of these rule changes directly affected operating reactor licensees.

- <u>Safety Significance Categorization Process</u>: This clarification notes that the NRC staff's endorsement of the safety significance categorization process described in NUMARC 93-01, Revision 2, is limited to the maintenance rule. It also notes that RG 1.160 is expected to be revised in the future, if needed, to reflect the proposed regulatory guide on the use of probabilistic risk assessment (PRA) in regulatory matters. Such a revision to RG 1.160 is desirable for reasons of regulatory consistency, and is also consistent with the NRC's intent (as presented in the statements of consideration (SOCs) for the maintenance rule) that licensees' approaches to the rule using PRA could change as a result of technological improvements and experience.

- <u>Scope of SSCs Under 10 CFR 50.65(b)</u>: Although requirements for the scope of SSCs under the rule defined in paragraph (b) are prescriptive, experience indicated the need for clarifications primarily related to the following two items:

  - <u>Could Cause</u>: This clarification gave guidance on how to identify the non-safety-related SSCs that should be considered within the scope because their failure *could cause* a reactor scram or safety system actuation.

  - <u>SSCs Relied Upon To Mitigate Accidents or Transients or Used in Emergency Operating Procedures (EOPs)</u>: This modification clarified that this criterion for SSCs within the scope of the rule includes SSCs that are used to directly address the accident or transient or are explicitly used in the EOPs, and provide a significant fraction of the mitigating function. In addition, this criterion includes SSCs that are *necessary* to support mitigation of accidents or transients or to facilitate the use of EOPs (even though the SSCs do not directly address the accidents or transients or may not be explicitly referenced in EOPs). The NRC staff added this clarification after finding that a number of licensees excluded communications and emergency lighting systems even though these systems are good examples of SSCs included under the rule for this criterion.

---

[5]Revision 1 to the RG 1.160 was issued in January 1995 to reflect an earlier rule change that extended the period of the paragraph (a)(3) periodic evaluations from at least once every 12 months to at least once per refueling cycle not to exceed 24 months between evaluations.

- **MPFFs as a Reliability Indicator:** As previously discussed, the staff has cited numerous licensees for not establishing reliability performance measures that are commensurate with safety because the licensees did not have a sound technical basis for the performance measures they established. Revision 2 to the RG clarified how MPFFs can be used as an indicator of reliability.

  In addition, the following guidance on reliability indicators may be proposed for RG 1.160, Revision 3: "Many licensees monitor both functional failures (FFs) and MPFFs under their MR monitoring program. The definition of a FF is an SSC that fails to perform its MR intended function; however, the FF may not be caused by maintenance but caused by other factors (e.g., operator error, design deficiency). Licensees may choose to monitor FFs each time period rather than MPFFs each time period since random failures resulting from FFs each time period versus MPFFs each time period may give more accurate information on equipment reliability. In addition, MPFFs each time period should be considered a subset of FFs each time period. Licensees are encouraged to establish conservatism within their monitoring programs; however, licensees and NRC inspectors should understand that monitoring FFs each time period is not required under the maintenance rule."

- **Structural Monitoring:** During the pilot program, it became apparent that there was a need to improve the structural monitoring guidance in NUMARC 93-01, Revision 1 (i.e., some additional guidance was presented in Revision 2 to NUMARC 93-01). As previously stated, many MRBIs identified structural monitoring as an inspection followup item until additional guidance was available. The industry developed additional guidance in NEI 96-03, but it had not been completed in time to reference it in Revision 2 to RG 1.160. The staff determined that it was important to provide additional guidance to licensees quickly; therefore, the NRC staff added guidance to RG 1.160, Section C, Regulatory Position 1.5, "Monitoring Structures."

- **Normally Operating SSCs of Low Safety Significance:** As previously stated, normally operating SSCs of low safety significance are typically monitored at the plant level (i.e., plant events that cause scrams, safety system actuations, or transients). Experience gained during the pilot program and baseline inspections indicated the need for additional clarification in this area. Accordingly, RG 1.160, Revision 2, clarified the treatment of normally operating SSCs of low safety significance. These clarifications were needed because the NRC staff noted during the MRBIs that some licensees only monitored the number of unplanned *automatic* scrams, and were not monitoring unplanned *manual* scrams even when the manual scram was initiated in anticipation of an automatic scram. Given that one of the principal reasons for developing the rule was the number of reactor scrams (*both manual and automatic*) caused by LSS normally operating SSC failures in the balance of plant, licensees should monitor *all* unplanned scrams in order to assess the effectiveness of their preventive maintenance for those SSCs monitored at the plant level. See Section 2.5.5 of this report for additional details on this issue.

Conclusions

After NRC issued RG 1.160, Revision 2, the NRC staff determined that adequate guidance to verify licensees efforts to implement the requirements of the maintenance rule existed; however, additional revisions will be needed to reflect 10 CFR 50.65(a)(4) rulemaking activities.

Recommendations

Revise RG 1.160 to enhance regulatory guidance on the basis of lessons learned from completing the MRBIs. In addition, revise RG 1.160 after rulemaking activities are completed on 10 CFR 50.65(a)(4) to *require* licensees to assess and manage increases in risk before performing maintenance activities.

2.2 Development of and Revisions to Inspection Procedure 62706

During the implementation period of the maintenance rule, it became apparent that the guidance documents would require revision to reflect lessons learned. That is, until full implementation of the rule is observed, it is difficult to determine whether the guidance documents and inspection procedures have sufficient details to ensure compliance with requirements. The need for this iterative process arises from the general requirements of the rule and the flexibility licensees are given to implement the requirements of the rule.

Findings

Given this need for an iterative process in the inspection guidance, the NRC staff issued IP 62706, Revision 1, dated December 31, 1997. This procedure referenced the latest version of MR guidance documents (i.e., NUMARC 93-01 and RG 1.160), which reflected lessons learned from earlier MRBIs. The procedure also discusses the inspection objectives, requirements, and guidance to verify implementation of each requirement of the rule.

Conclusion

The NRC staff concluded that IP 62706 will need to be revised again to include NRC approved rulemaking on 10 CFR 50.65(a)(4), which will require licensees to perform risk assessments before taking equipment out of service for maintenance.

Recommendation

Revise IP 62706 to incorporate new inspection guidance, including Commission approved rulemaking on 10 CFR 50.65(a)(4).

2.3 Scope of SSCs Within 10 CFR 50.65(b)

The scope of SSCs that are required to be maintained within the rule includes safety-related SSCs that are relied on to function during and following design-basis events to ensure the integrity of the reactor coolant pressure boundary, the capability to shutdown the reactor and maintain it in a safe shutdown condition, or the capability to prevent or mitigate the consequences of accidents that could result in potential offsite exposures comparable to the limits defined in 10 CFR Part 100. In addition, the required scope includes non-safety-related SSCs that meet any one of the following criteria: (1) relied upon to mitigate accidents or transients or are used in plant emergency operating procedures (EOPs); (2) failure could prevent a safety-related SSC from fulfilling its intended function; or (3) failure could cause a reactor scram or actuation of a safety-related system.

In order to increase effectiveness and efficiency, the NRC inspection teams requested advance copies of processes and procedures that the licensee used to determine which SSCs would be included in the scope of the rule. The teams also requested a list of SSCs in scope as well as SSCs not in scope. The teams then focused attention on SSCs that were excluded from the scope of the licensee's MR program. If a particular SSC appeared to fall within the scope of the rule but was excluded from the licensee's MR program, the teams determined whether the licensee had an adequate technical basis for excluding the given SSC. The teams also reviewed the plant's final safety analysis report (FSAR), EOPs, and the individual plant evaluation (IPE) to determine if the licensee made proper decisions concerning the scope of SSCs and their functions that were under the rule.

Findings

To determine which SSCs should be within the scope of the MR, licensees typically reviewed the plant FSAR, EOPs, and the IPE. The licensees also used other documents such as plant equipment lists and quality assurance (QA) lists to identify SSCs that should be within the scope of the rule. Some licensees used their expert panel and most licensees reviewed industry-wide operating experience to make decisions on scoping SSCs. In addition, most licensees evaluated system functions and presented this information in MR system basis documents to identify all the components within the scope of the MR for a particular system. Identifying system functions also made it easier for licensees to identify MR FFs and MPFFs when system equipment failed.

The MRBI teams found that differences in licensees' programs and differences in plant design (e.g., system boundaries), even among plants that have similar nuclear steam supply systems (NSSSs), can result in significant differences in the numbers and types of SSCs included within the scope of the rule. For example, many licensees scope system functions in addition to individual SSCs and individual system boundaries against the criteria under paragraph 50.65(b). Therefore, the industry and NRC inspectors had to be very careful when comparing one plant to another when evaluating MR decisions on the scope of SSCs under 50.65(b) at a particular plant.

Licensees generally used NUMARC 93-01, Section 8, "Methodology To Select Plant Structures, Systems and Components," to make decisions on SSCs that should be within the scope of the rule. As determined during the MRBIs, licensees identified most or all of the SSCs that should be included within the scope of the rule. However, the MRBI teams did identify plants that made inadequate decisions on a few safety-related and some non-safety-related SSCs within the scope of the rule. The MRBI teams identified an average of two to five additional SSCs or SSC functions that should have been included in the rule. Given these results, the NRC staff concluded that the industry generally did a good job of identifying SSCs required to be within the scope of the rule.

As a result of the MRBIs or as identified by the licensee's self-assessment process, the NRC staff identified issues involving SSCs that were either not included within the scope of the rule at the time the rule became effective, or were not appropriately classified as being within the scope of the licensee's MR program at the time of the inspection. These were cited as violations in the associated MRBI reports with enforcement actions taken. Examples of structures that were not included within the scope of the rule but that should have been are the following: the switchyard relay building, transformer pads, circulating water intake and discharge bays, building tunnels, the turbine building, and various sumps. Examples of systems and components that were not included within the scope of the rule are air compressors, accident sampling systems, chilled water systems, circulating water traveling screens, communications equipment, condenser air removal systems, control rod position indication, control room annunciators, freeze protection equipment, fuel handling cranes, electrical cable, emergency lighting, main generator support systems, non-vital electrical power distribution systems, plant computer, rod block monitor, service air, service water, transformers, ventilation support systems, and various radiation monitors. Examples of system functions that were erroneously not included within the scope of the MR are some safety-related functions, accident mitigating fire protection functions, demineralized water function to fill the condensate storage tanks, emergency raw cooling water functions, rod control functions, containment air handing functions, a feedwater recovery function, and other functions needed in technical specification operating modes 2 through 5 for a boiling-water reactor (BWR) plant (i.e., startup, hot shutdown, cold shutdown, and refueling).

As indicated during the MRBIs, some SSCs or SSC functions can be excluded from the scope of the rule provided the licensee has an adequate technical basis. For example, at one plant, the team determined that the cooling tower of the circulating water system should be within the scope of the rule since a site-specific failure of a cooling tower function led to a reactor trip event. At other plants, no site-specific events occurred involving a cooling tower, industry-wide operating experience did not identify a similar failure mode, and the licensee provided an adequate technical basis for excluding the cooling tower.

Conclusions

Generally, licensees did a good job of identifying SSCs within the scope of the rule. On average, the MRBI teams identified only two to five additional SSCs or SSC functions that should have been included within the scope of the rule for each site.

Since the maintenance rule is a performance-based regulation, licensees have the flexibility to add or remove SSCs from the scope of 10 CFR 50.659(b) if an adequate technical basis exists for including or excluding the SSC in question. In addition, licensees can exclude a non-safety-related SSC if its performance demonstrates that it did not contribute to events that require non-safety-related SSCs to be included within the scope as defined under 10 CFR 50.65(b)(2).

Recommendation

Using the flexibility of rule implementation guidance, licensees should provide adequate technical bases for making performance based decisions on the scope of SSCs under 10 CFR 50.65(b)

2.4 Risk Determination Process—Classifying SSCs as High or Low Risk Significant

The MRBI teams reviewed the programs and procedures that licensees used to implement the risk determination guidance in NUMARC 93-01, Section 9.3.1, "Establishing Risk Significant Criteria." The NUMARC 93-01 guidance for determining risk significance recommended that an expert panel should use the Delphi method of NUREG/CR-5424, supplemented by PRA or IPE insights, to categorize SSCs as HSS or LSS SSCs within the scope of the rule. The guidance also uses PRA importance measures such as risk reduction worth (RRW), risk achievement worth (RAW), and core damage frequency (CDF) contribution and provides this information to the expert panel when making risk ranking decisions for SSCs within the scope of the rule. In addition, NUMARC 93-01, Section 9.3.1, provided numerical thresholds for these importance measures (e.g., RAW>2, RRW>1.005, and cut sets that account for 90% of the CDF) as a set of quantitative criteria for identifying HSS and LSS SSCs.

Findings

Results of MRBIs indicated that all licensees followed the guidance in NUMARC 93-01 and used the prescribed importance measure threshold limits to determine HSS and LSS SSCs within the scope of the rule. Also, some licensees used alternative importance measures (e.g., Birnbaum importance, Fussel-Vesely importance) in the risk ranking of HSS and LSS SSCs. In these cases, the threshold limits of the alternative importance measures were equivalent to the criteria prescribed for the more commonly used importance measures. However, the NRC staff noted that the universal values of RAW, RRW, and other importance measures prescribed in NUMARC 93-01 do not reflect the plant-specific risk profiles on a case-by-case basis. Therefore, the use of these values alone could cause inconsistencies in the risk significance determination of SSCs within the scope of the rule.

To compensate for potentially inconsistent risk ranking of SSCs in the PRA from the use of thresholds for the selected importance measures, all licensees used the expert panel process to determine the final list of HSS and LSS SSCs within the scope of the rule. Typically, the expert panel consisted of personnel experienced in plant operations, maintenance, and engineering, along with an individual with expertise in PRA. In many cases, the expert panel was also responsible for determining which SSCs should be within the scope of the MR, moving SSCs from paragraph (a)(2) to (a)(1) and from (a)(1) back to (a)(2) on the basis of SSC performance, identifying SSCs that should have goals or performance measures established, recommending corrective actions needed to improve SSC performance, or providing insights on other elements of the MR program.

2.4.1 Use of PRA Importance Measures in Determining Risk-Significant SSCs

NUMARC 93-01 discusses the common methods used to group SSCs on the basis of safety (risk) significance. An acceptable approach for determining risk significance is to use PRA importance measures, which provide the quantitative measure of relative risk impact (on the basis of a risk measurement such as CDF) of the individual SSC if the SSC's reliability or availability is decreased. This approach also requires the expert panel to consider issues related to design and operating experience in order to appropriately rank SSCs and to avoid mis-categorizing HSS SSCs into the LSS group.

All licensees used PRA importance measures based on Level 1 PRA analysis (e.g., CDF cut set contribution, RAW, RRW) to support quantitative decisions regarding the safety significance of SSCs within the scope of the rule. In almost all cases, if any SSC importance measures exceeded the NUMARC 93-01 thresholds, licensees classified the SSC as HSS. At some sites, licensees did not utilize their Level II PRA and/or shutdown PRA information to address the safety-significance determination of SSCs that were not modeled in Level I PRAs (e.g., containment isolation valves) since importance measure information for large early release frequency (LERF) was not submitted to the expert panel. However, in most cases, the licensees' expert panels took into account operating experience and design-basis information to appropriately identify individual SSCs that would be categorized as HSS SSCs in Level II PRAs.

During the MRBIs, the teams reviewed plant-specific information to determine whether the basic data assumptions used in the PRA were consistent with the actual operating experience data. Overly conservative assumptions can elevate the importance of certain SSCs while masking the true importance of others. For example, the use of generic reliability and maintenance unavailability data for certain SSCs, which may be more conservative than plant-specific values, can skew the risk-ranking results. Technical issues associated with using PRA importance measures in other risk-informed applications are addressed in Appendix A to NRC Regulatory Guide 1.174, "An Approach for Using Probabilistic Risk Assessment in Risk-Informed Decisions on Plant-Specific Changes to the Licensing Basis," dated July 1998.

The scope of the rule extends to a variety of SSCs that may not be modeled in sufficient detail in the PRA to support decisions regarding safety-significance determinations. Many complex systems are commonly modeled as "super components" or "black boxes" in PRA studies (i.e., diesel generators, turbine trip systems, etc.). In several PRAs, the risk impact of system performance may only be modeled via an *initiating event* (i.e., loss of turbine cooling water, loss of instrument air, etc.). This artifact of PRA modeling might result in inappropriate ranking of certain systems because their importance could be obscured by the direct or indirect contributions to risk (e.g., common-cause failures) of the system components. For these particular SSCs, the expert panel process should be able to qualitatively assess the risk contribution of the SSC and appropriately rank the SSC.

Another factor that affects risk-ranking results from a PRA is the modeling of operator actions in dominant accident sequences. Typically, subjective judgment is involved in estimating the human error probabilities (HEPs) associated with operator actions modeled in the accident sequence cut sets.[6] In some situations, very high success probabilities (i.e., low HEPs) for operator recovery actions are assigned to a sequence, and this results in related SSCs being ranked as low risk contributors. Thus, it is not desirable to determine the safety-significance of SSCs affected by operator recovery actions that are only modeled in dominant accident scenarios. In the discussions regarding risk-significance determination methods using the CDF contribution and RRW measures, NUMARC 93-01 recommends eliminating cut sets that are not related to maintenance (e.g., cut sets containing only operator recovery errors, and external or initiating events) to avoid the potential "masking" effects due to operator recovery actions, and external or internal initiating events. In these situations, the removal of such cut sets should be done cautiously to ensure that the eliminated cut sets do not implicitly account for SSC maintenance activities. Otherwise, MPFFs caused by operator errors could be screened out from the risk-ranking process. Thus, the treatment of operator actions in licensees' PRA models and their effects on the risk-ranking results should be presented to the expert panels to support the appropriate ranking of the SSCs.

---

[6]A cut set is a set of elements or components in a system whose failure will cause the system to fail. For example, a minimal cut set is the smallest set of failed components in a system that causes the system to fail. In practice, reliability analysts use cut sets to identify different combinations of failed components that cause a system to fail.

2.4.2 Use of the Expert Panel in Determining Risk-Significant SSCs

During the MRBIs, the inspection teams found that licensees used a multi-disciplinary expert panel review process with experts in PRA, operations, engineering, and maintenance. This enabled licensee personnel to collectively make sound risk-informed decisions on HSS and LSS SSCs within the scope of the rule. During early MRBIs, the teams found some expert panels with weak PRA knowledge. In most cases, the expert panel made conservative decisions using other risk-informed approaches (e.g., large early release frequency (LERF) shutdown PRAs) in determining HSS and LSS SSCs within the scope of the rule. In a few cases, licensees' expert panels did not consider risk measures from other risk-informed analyses, and this led to non-conservative decisions.

To determine HSS and LSS SSCs within the scope of the rule, the expert panels reviewed PRA and non-PRA information needed to integrate the risk determination decisions through multi-disciplinary reviews of PRA importance measures, operational experience, engineering judgment, and maintenance results. However, some licensees utilized different approaches to integrate the information when making risk ranking decisions. For example, some licensees used a horizontal and vertical slice two-dimensional graph of Fussel-Vesely importance versus Birnbaum importance to determine HSS and LSS SSCs. The threshold values were equivalent to those used for RAW and RRW found in NUMARC 93-01, Section 9.3.1, "Establishing Risk Significant Criteria." The results could be plotted on a graph providing a simple two- dimensional plot with four quadrants in which SSCs were identified as HSS and LSS based on a system's importance measure values.

2.4.3 Use of PRA and the Expert Panel in Other Regulatory Applications

The risk significance determination process described in NUMARC 93-01 is generally adequate for grouping the SSCs into two safety-significance categories for the implementation of maintenance rule requirements. However, this process may not be adequate for other regulatory applications, such as graded quality assurance, in-service testing, in-service inspection, and plant-specific changes to the licensing bases. Each of these applications may require additional consideration of issues that uniquely affect the risk ranking of SSCs for that particular application. For example, the risk ranking of SSCs for a graded quality assurance program would recommend that safety-significance categorization criteria take into account the plant-specific baseline risk level.

The use of a multi-disciplinary expert panel in MR implementation ensures proper integration of deterministic and probabilistic insights to support decisions regarding the safety significance of SSCs. Although other risk-informed regulatory applications may benefit from using the multi-disciplinary expert panel approach, details regarding the makeup of the expert panel may vary, as will the manner in which probabilistic and deterministic considerations are integrated for use in a particular application. In addition, many licensees used expert panels to support decisionmaking activities associated with other parts of the rule, such as scope of the rule, level of monitoring, and paragraph (a)(1) versus (a)(2) determinations.

Conclusions

The methods licensees used to establish risk significance were consistent with the guidance in NUMARC 93-01. The use of an expert panel review process to integrate deterministic considerations with PRA or IPE risk insights is an appropriate and practical method of determining SSC risk significance. During early MRBIs, the teams found some expert panels with weak knowledge of PRA. At most sites, the expert panel members were knowledgeable of the MR and had extensive industry experience to support the decisionmaking process for determining the safety significance of SSCs. The required composition of the expert panels was also consistent with the guidance in NUMARC 93-01.

Recommendations

As a result of the MRBIs, the NRC staff identified the following issues and recommendations concerning the risk significance determination process. These recommendations should be considered when clarifying the related guidance in NUMARC 93-01:

*   NUMARC 93-01, Sections 9.3.1.1 and 9.3.1.2, recommended that licensees eliminate RRW measures and cut sets that are not specifically related to maintenance (e.g., operator error, external and initiating events) from the risk determination process. This guidance was presented to avoid the potential masking of the importance of certain SSCs by PRA modeling assumptions (e.g., higher probability estimates) used for operator errors, and external and internal events. In a few cases, this guidance caused licensees to identify some SSCs as LSS rather than HSS. RRW measures and cut sets should be removed cautiously to ensure that the eliminated cut sets do not implicitly account for maintenance activities associated with certain SSCs that have high importance ranking.

*   Plant-specific PRA models used for risk-importance analyses should be of sufficient quality to ensure consistent results in the SSC safety-significance categorizations. In determining the safety significance of SSCs, all licensees performed importance analyses using PRA models that were developed for IPEs or individual plant examination of external events (IPEEEs). The quality of the PRAs used at most of the licensee sites has not been assessed through peer or industry reviews. For MRBI purposes, the PRA was assumed to have sufficient quality to support the risk categorization process since there was an expert panel process in place to compensate for limitations in the PRA. This issue of PRA quality is addressed in ongoing NRC and industry initiatives to identify requirements for PRA standards in a PRA certification process.

*   For future risk-informed, performance-based inspection initiatives, the licensee's risk-determination methodologies for determining HSS and LSS SSCs within the scope of other regulatory applications should also take into account the plant-specific baseline risk level. This could result in more suitable importance measure thresholds being used for the specific plant and regulatory applications under review.

*   NRC and licensee resource commitments for risk-informed, performance-based regulatory activities are initially high, and should be acknowledged and committed to up front. Risk-

informed, performance-based regulatory activities require coordination with the industry to develop explicit implementation guidance.

- The risk ranking results should be re-evaluated whenever major plant design changes are implemented, the PRA models are updated, and new reliability and availability data become available.

- Currently, NUMARC 93-01, Section 9.3.1, provides guidance only on the use of an expert panel process to determine the safety significance of SSCs by evaluating PRA risk ranking results in conjunction with the design and operating experience considerations. The industry should consider using the expert panel to support other decision making activities associated with other parts of the rule such as the scope of the MR, establishment of goals and performance measures, when SSCs should be moved from paragraph (a)(2) to (a)(1) or from paragraph (a)(1) back to (a)(2), corrective actions to improve SSC performance, the periodic evaluations under paragraph (a)(3) of the rule, and performing safety assessments before conducting maintenance activities under the new paragraph (a)(4).

2.5 Goal Setting, Monitoring, and Preventive Maintenance

The MRBI teams reviewed licensees' MR program documents, procedures, and records at each site to evaluate the established process to set goals and monitor SSC performance in accordance with paragraph (a)(1) of the rule, and to verify preventive maintenance was effective in accordance with paragraph (a)(2) of the rule. At each site, the teams selected a sample of SSCs that were categorized as HSS and LSS and were monitored under paragraph (a)(1) or tracked under paragraph (a)(2) for a detailed vertical slice sample reviews[7]. In addition, the teams performed horizontal slice program reviews[8] of SSCs monitored under paragraph (a)(1) or tracked under paragraph (a)(2) to verify that adequate goals and performance measures were respectively established.

2.5.1 Categorizing SSCs Under Paragraph (a)(1) or (a)(2)

NUMARC 93-01, as endorsed by RG 1.160, states that SSCs are subject to goal setting and monitoring in accordance with paragraph (a)(1) of the rule whenever performance criteria are exceeded or repetitive MPFFs occur. The MRBI teams verified that licensees appropriately followed this guidance when determining whether SSCs should be monitored under paragraph (a)(1) or tracked under paragraph (a)(2). The teams also verified that licensees were adequately identifying MPFFs or unavailability data, and moving SSCs to the (a)(1) monitoring category when goals or performance measures were exceeded. When SSCs exceed their goals or performance measures, licensees should place SSCs into the paragraph (a)(1) monitoring category, establish more effective monitoring goals, and complete root cause or cause

---

[7]The term "vertical slice sample reviews" means that the inspector verified that the licensee implemented all MR requirements for a sample of SSCs under the scope of the maintenance rule.

[8]The term "horizontal slice program reviews" essentially means that the inspector verified that the licensee implemented a particular MR requirement for all SSCs under the scope of the rule.

determination analyses as well as corrective actions needed to improve overall SSC performance. As stated in RG 1.160, Revision 2, the number of SSCs in the paragraph (a)(1) monitoring category will not be used by the NRC as an indicator that licensees' preventive maintenance programs are ineffective. Since the MR is a risk-informed, performance-based regulation, licensees are given flexibility with their monitoring programs to establish the most effective means of improving overall SSC performance.

The MRBI teams reviewed a sample of HSS and LSS SSCs evaluated or tracked[9] under paragraph (a)(2) of the rule to determine if they should be categorized under paragraph (a)(1), where they would be subject to goal setting and monitoring. The teams also verified that SSCs categorized under paragraph (a)(1) had goals established that took into account root cause or cause determination analyses and corrective actions needed to improve SSC performance. In addition, the teams verified that licensees took industry operating experience into account when establishing goals.

Findings

The MRBI teams determined that many licensees did not in some instances (1) establish adequate goals commensurate with safety, (2) provide an adequate technical basis for performance measures, or (3) statistically link these goals and performance measures to the reliability and availability assumptions in the PRA. Another common deficiency was that licensees did not adequately monitor or keep track of FFs, MPFFs, repeat MPFFs, and/or unavailability for some HSS and LSS standby SSCs. In addition, some licensees' root cause or cause determination analyses and corrective actions were inadequate and this deficiency resulted in inadequate evaluation of SSC performance. In particular, the MRBI teams attributed licensees' inadequate monitoring to the following causes:

- limited human resources;

- inadequate tracking and trending of data in databases used to monitor and trend equipment performance;

- inadequate training of system engineers responsible for evaluating and tracking FFs, MPFFs, repetitive MPFFs, and unavailability data and entering it into a database; and

- inadequate evaluation of PRA risk insights when establishing goals and performance measures.

During early MRBIs, licensees placed only a small number of SSCs into the paragraph (a)(1) monitoring category. Many licensees appeared reluctant to move SSCs into the paragraph (a)(1) monitoring category since their perception was that the NRC and licensee management would view numerous SSCs in paragraph (a)(1) as an indicator of ineffective maintenance. The MRBI teams also found a number of SSCs that exceeded their performance measures or had inadequate performance measures. During the course of the MRBIs, the teams reminded licensees that a reluctance to place SSCs into paragraph (a)(1) when performance measures are exceeded or repetitive MPFFs occur constitutes a potential violation of

---

[9]For the purpose of this document, the term "evaluated" or "tracked" will be used to indicate the process licensees use to demonstrate the effectiveness of preventive maintenance for SSCs monitored under paragraph (a)(2) of the rule.

the MR; therefore, licensees should move SSCs into paragraph (a)(1) when they identify maintenance related performance problems. The MRBI teams also continued to stress to licensee representatives that the NRC staff would not consider the number of SSCs in the paragraph (a)(1) monitoring category as an indicator of ineffective maintenance. As MRBIs continued, more licensees used the paragraph (a)(1) monitoring category to focus management attention on SSCs with maintenance problems that needed corrective actions and improved performance.

Conclusions

In some instances, licensees did not (1) establish adequate goals commensurate with safety, (2) provide adequate technical bases for performance measures that demonstrate effective preventive maintenance, and/or (3) statistically link goals or performance measures or both to the assumptions in the PRA. Several licensees also did not adequately monitor or track FFs, MPFFs, repetitive MPFFs, and/or unavailability for some HSS and LSS standby SSCs. Some licensees' root cause or cause determination analyses and corrective actions were inadequate. The MRBI teams concluded that some inadequate monitoring was caused, in part, by a licensee's reluctance to move SSCs into the (a)(1) monitoring category or by a licensee's lack of awareness of goals or performance measures that were insufficient or exceeded.

The perceived reluctance of licensees to place SSCs in the paragraph (a)(1) monitoring category was effectively addressed through guidance presented in RG 1.160, Revision 2, dated March 1997, and Information Notice (IN) 97-18, "Problems Identified During Maintenance Rule Baseline Inspections," dated April 14, 1997.

Recommendations

Licensees may consider the number of SSCs in the paragraph (a)(1) monitoring category as an internal indicator of the efficacy of the MR at their plant. The NRC does not consider the number of SSCs in the (a)(1) monitoring category to be an indicator of ineffective maintenance. The NRC may consider the number of SSCs with repetitive returns to the (a)(1) monitoring category as an indicator of MR efficacy.

2.5.2 Monitoring SSCs in Accordance With Paragraph (a)(1) and Taking Corrective Actions

The NRC reviewed licensees' root cause or cause determination processes, corrective actions taken, and goals established to improve overall SSC performance for HSS SSCs, LSS standby SSCs, and LSS normally operating SSCs monitored in accordance with paragraph (a)(1) of the rule. In particular, the NRC evaluated whether licensees established adequate (1) reliability and availability goals for HSS SSCs; (2) reliability, availability, or condition monitoring goals for LSS standby SSCs; and (3) plant level goals (e.g., scrams, safety system actuations, or unplanned capacity lost factors) for LSS normally operating SSCs.

Findings

Most licensees used the requirements of paragraph (a)(1) to develop site-specific MR system health plans that established monitoring goals, completed root cause and/or cause determination analyses, and implemented corrective actions needed to improve overall SSC performance.

During early inspections, the MRBI teams determined that some licensees did not establish goals commensurate with safety in all cases. For example, a number of licensees established a goal for reliability (e.g., number of MPFFs in a time period) without an adequate technical basis. In some cases, when compared to the number of demands on a particular SSC, the number of MPFFs allowed would indicate a much lower reliability than the licensee assumed in its risk analysis which was used to determine an SSC's risk significance. More specifically, licensees did not demonstrate that the goals or performance measures or both preserved the assumptions defined by the plant-specific PRA, IPE, or other risk determining analyses. Also, some licensees did not establish an availability goal or performance measures for certain HSS SSCs. Licensees' corrective actions included completing sensitivity studies using importance measures (e.g., RAW) to determine more appropriate goals linked to the assumptions in the PRA for SSCs being monitored under paragraph (a)(1).

In addition, paragraph (a)(3) of the rule requires balancing the reliability achieved through preventive maintenance activities against the objectives of minimizing unavailability. When establishing goals under paragraph (a)(1) of the rule, some licensees could not adequately perform the balancing because the reliability goals for some SSCs were inadequate or SSC availability goals were not established.

Some licensees were late with implementing their own QA audit findings and training programs for system engineers who would implement the monitoring and trending of FFs, MPFFs, repetitive MPFFs, and excessive unavailability. This included training programs to evaluate root cause and cause determination of FF, MPFF, repetitive MPFFs, and large unavailability; and taking corrective actions to establish goals to improve overall SSC performance.

In addition, the MRBI teams found that some licensees monitored components failures (i.e., breakers, motorized operated valve (MOV) actuators, solenoid-operated valves, limit switches, relays, etc) at the system or train level when functional failures of these components occurred across several different and diverse systems or system trains. In many instances, the systems would not exceed their performance criteria; however, a history of common-cause failures at the component level would necessitate evaluating the component for monitoring in accordance with paragraph (a)(1). To address this situation, many licensees categorized similar components as a separate class of components across system or train boundaries under paragraph (a)(1) when repetitive MPFFs occurred and took corrective actions to eliminate the common-cause failure. However, some licensees did not monitor these similar components under paragraph (a)(1) because system or system train level performance was acceptable and common-cause repetitive MPFFs were not identified.

Conclusions

The overall results of the MRBIs indicated that, in general, licensees adequately moved SSCs to the paragraph (a)(1) monitoring category when performance measures were exceeded or repeat MPFFs occurred. However, during early MRBI inspections, many licensees in some instances did not establish goals commensurate with safety or did not adequately monitor SSC performance. In addition, many licensees did not always establish adequate performance measures that would demonstrate that SSCs were being maintained by effective preventive maintenance; therefore, the technical basis for monitoring these SSCs under paragraph (a)(2) was flawed and these SSCs should have been evaluated for monitoring under paragraph (a)(1) as discussed in Section 2.5.3 below.

Recommendations

Even if individual system or system train level performance is acceptable, licensees should monitor similar components (i.e., circuit breakers, MOVs, solenoid-operated valves, limit switches, relays, etc) across system or train boundaries in accordance with paragraph (a)(1) when common-cause repetitive MPFFs occur.

2.5.3 Demonstrating the Effectiveness of Preventive Maintenance for SSCs in Accordance With Paragraph (a)(2)

The rule in paragraph (a)(2) states, in part, that "monitoring as specified in paragraph (a)(1) is not required where it has been demonstrated that the performance or condition of SSCs is being effectively controlled through the performance of appropriate preventive maintenance, such that the SSC remains capable of performing its intended function." NUMARC 93-01, Section 9.3.2, "Performance Criteria for Evaluating SSCs," established industry guidance for determining if the performance or condition of SSCs is being effectively controlled with adequate preventive maintenance programs.

As described in NUMARC 93-01, performance criteria that are used to demonstrate effective preventive maintenance for HSS SSCs should be established to ensure that reliability and availability assumptions used in the plant-specific safety analysis are maintained or adjusted. Performance criteria for LSS standby SSCs can be reliability, availability, or condition, but, at a minimum, it should include reliability. Performance criteria for LSS normally operating SSCs should be evaluated against plant level performance criteria (e.g., causes scrams, unplanned capability loss factor, or failure of safety-related SSCs). The team performed both horizontal slice program reviews and vertical slice sample reviews for SSCs tracked in accordance with paragraph (a)(2) of the rule.

Findings

The results of the MRBIs generally indicated that most licensees followed the industry guidance to implement paragraph (a)(2) of the rule. NUMARC 93-01 calls for licensees to establish and evaluate performance criteria and to verify that the performance criteria have not been exceeded for SSCs in paragraph (a)(2). Within this area, the MRBI teams noted similar types of deficiencies throughout the industry. In particular, licensees did not always establish appropriate (1) performance criteria commensurate with safety and linked to the reliability or availability assumptions in the PRA for some HSS SSCs, (2) performance criteria (i.e., reliability, availability or condition) for some LSS standby SSCs, and (3) plant level performance criteria for most balance of plant systems in the turbine building. In addition, a few licensees did not establish adequate specific performance measures for SSCs that perform functions that protect the fuel cladding during safe storage of nuclear fuel in the spent fuel pool. In many cases, these performance measures would not demonstrate effective preventive maintenance; therefore, the technical basis for evaluating these SSCs in accordance with paragraph (a)(2) was flawed and these SSCs should have been considered for monitoring under paragraph (a)(1).

A few licensees took exceptions to the guidance contained in NUMARC 93-01 by establishing new and different performance measures. The flexibility of the MR allows licensees to establish different methods to evaluate SSCs in accordance with paragraph (a)(2) as long as the new method adequately assesses the effectiveness of preventive maintenance activities. For example, one plant established a new performance measure, called maintenance preventable redundancy functional failure (MPRFF), to track the reliability of systems with redundant trains. MPRFFs occur when spare pumps are unavailable or fail as a result of maintenance related activities. One licensee's general service water (GSW) system contained five pump trains of which any combination from two to five pump trains could perform the intended function of the system depending on the temperature and level conditions of its heat sink. The licensee determined that at certain times of the year there was no safety benefit associated with evaluating unavailability or failures of redundant pump trains; however, tracking loss of redundancy was a good indicator of the capacity of the system under degraded conditions and, therefore, offered some safety benefit. The NRC determined that this was acceptable, given the specific circumstances and controls for the affected system function.

Conclusions

In general, the MRBI teams found that the industry had developed adequate performance criteria for evaluating SSCs in accordance with paragraph (a)(2) of the rule; However, in many cases, the performance criteria developed would not demonstrate effective preventive maintenance; therefore, the technical basis for evaluating these SSCs in accordance with paragraph (a)(2) was flawed and these SSCs should have been considered for monitoring under paragraph (a)(1).

Recommendations

The industry may wish to consider additional methods (e.g., MPRFFs) to demonstrate the effectiveness of maintenance on redundant equipment (i.e., instrument and service air compressors, service water pumps, etc.). This could eliminate the need to track unavailability of equipment trains that are removed from service for extended periods of time.

2.5.4 Safety Considerations in Establishing Goals or Performance Criteria

The rule in paragraph (a)(1) requires licensees to monitor the performance or condition of SSCs against licensee established goals in a manner sufficient to provide reasonable assurance that such SSCs as defined in 50.65(b) are capable of performing their intended function. Licensees must establish goals commensurate with safety. When the performance or condition of an SSC does not meet established goals, the licensee must take appropriate corrective actions to improve SSC performance.

Monitoring as specified in paragraph (a)(1) is not required in paragraph (a)(2) if performance measures established for SSCs demonstrate that the SSCs are maintained by an effective preventive maintenance program so that the SSCs remain capable of performing their intended function.

Findings

Each licensee performed a risk determination for SSCs within the scope of the rule at their own facilities. These risk determinations establish the basis for considering safety when setting goals or performance measures under paragraph (a)(1) or (a)(2) of the rule. Licensees used the results of these risk determinations to decide whether goals or performance measures would be at the plant, system, train, component, or parameter level. Accordingly, system, train, or component level goals or performance measures were established for most HSS SSCs and LSS standby SSCs. A few very HSS SSCs (e.g., reactor protection system (RPS)) would require parameter or condition monitoring (e.g., channel level monitoring or component condition monitoring) to verify that certain components were not in degraded conditions before failures occurred (see Section 2.5.5 of this report for additional details). Condition monitoring would also be used for most structures (see Section 2.5.6 of this report for additional details). Plant level goals or performance measures were then established for the remaining LSS normally operating SSCs. The MRBI teams found that most licensees had considered safety throughout this process of determining risk and evaluating whether the goals or performance measures should be set at the plant, system, train, component, or parameter level.

Almost all licensees used an expert panel and information from the risk determination process to establish goals or performance measures at the plant, system, train, or component level, and in some cases, the parameter level. In addition, licensees performed historical system reviews for HSS and LSS standby SSCs. Since precise unavailability data were not readily available from plant records, the expert panel used its collective judgments to determine system unavailability data from monthly operating reports. To validate the unavailability data, licensees typically recalculated the PRA using the new unavailability values and confirmed that the results were consistent with the assumed unavailability values from the original PRA. Reliability data were easier to obtain by simply counting functional failures per number of demands. Most licensees selected goals and performance measures based on reliability and unavailability data assumed in the plant specific PRA. However, as determined during early MRBIs, some licensees did not adequately consider the reliability and availability assumptions found in their PRAs.

Conclusions

The results of the MRBIs indicated that all licensees considered safety throughout the risk determination process to evaluate whether goals or performance measures should be established at the plant, system, train, component, or parameter level. However, some licensees did not adequately consider the reliability and availability assumptions found in their PRAs.

Recommendations

None

2.5.5 Monitoring and Trending the Performance of Systems and Components

In the SOC, dated July 10, 1991, for this rule (NRC, 1991), the Commission stated that "where failures are likely to cause loss of an intended function, monitoring under paragraph (a)(1) should be predictive in nature, providing early warning of degradation." NUMARC 93-01 gives guidance for using predictive maintenance, inspection, testing, and performance trending for monitoring the performance or condition of SSCs in paragraph (a)(2) of the rule. In addition, train level monitoring provides a method to address degraded performance of a single train even though the system function is still available. When component level monitoring is determined to be necessary, it should be based on the component's contribution to a system function not meeting its performance measure or a system level goal.

Findings

In accordance with the guidance contained in NUMARC 93-01, as endorsed by RG 1.160, Revision 2, the MRBI teams found that nearly all licensees conducted an appropriate degree of system, train, and component level monitoring, which met the requirements of the rule based on the safety significance of the SSCs in question. Licensees monitored the reliability and availability of most HSS SSCs at the component train level for components that cause a train function to fail. In addition, most licensees monitored the reliability, availability, and/or condition of LSS standby SSCs at the train level where appropriate. In addition, the MRBI teams noted that most licensees monitored LSS normally operating SSCs at the appropriate plant level performance measures (i.e., causes scrams, unplanned capability loss factors, and/or safety system actuations).

In a few instances, licensees did not establish the appropriate degree of monitoring as described in NUMARC 93-01. The causes for this inadequate monitoring primarily involved (1) limited human resources; (2) inadequate tracking and trending of data in databases used to monitor and trend equipment performance; (3) inadequate training of system engineers responsible for evaluating and tracking FFs, MPFFs, repetitive MPFFs, and unavailability data and entering it into a database; and (4) inadequate evaluation of PRA risk insights when establishing goals and performance measures at the appropriate level (i.e., system, train, component).

The MRBIs also identified one common finding concerning the establishment of condition monitoring goals or performance measures for the reactor protection system (RPS) and engineered safety features actuation system (ESFAS). Current technical specification (TS) monitoring programs require periodic surveillance testing of RPS and ESFAS at the channel level. The MRBI teams maintained that licensees should monitor unavailability or failures or both at the train or channel level as an indicator of effective preventive maintenance on these instrumentation and control (I&C) systems. The MRBI teams determined that this form of SSC condition monitoring is already required by the TSs.

However, many licensees asserted that unavailability and/or failure monitoring at the channel level should not be required since a channel failure does not cause a failure of the system function. This approach was predicated on the assumption that once a channel fails, it is placed in the tripped or bypass condition so that the RPS can still perform its intended function.

The NRC determined that once a licensee discovers that a channel has failed, the TSs require licensees to reduce the channel trip logic (i.e., one out of two taken twice now becomes one out of two). Thus, the trip logic is reduced and subject to common cause failures that could trip the plant and cause challenges to safety functions. Therefore, at least train (e.g., half scrams) and/or channel level monitoring is desirable since channel maintenance trending information could give licensees indications of an impending problem. The RPS is also a HSS system in which many PRAs show that the probability of failure given the expected number of system demand failures each year is very small (e.g., RPS reliability is in the 10^{-5} range/year). Additional information on RPS reliability can be found in NUREG/CR-5500, Volume 2, "Reliability Study, Westinghouse Reactor Protective System, 1984-1995," dated April 1999 and NUREG/CR-5500, Volume 3, "Reliability Study, General Electric Reactor Protective System, 1984-1995," dated February 1999. Therefore, condition monitoring at the train (e.g., half scrams) and/or channel level could preclude system level failures associated with an anticipated transient without scram (ATWS). It is also a more desirable approach for determining the effectiveness of maintenance on the RPS and ESFAS. The MRBI teams also determined that licensees should take credit for existing monitoring programs under the TSs to implement the monitoring requirements of the rule.

In addition, the MRBI teams noted a weakness in NUMARC 93-01, Section 9.3.2, "Performance Criteria for Evaluating SSCs," at the plant level. The guidance implies that licensees should count MPFFs only for unplanned *automatic* reactor scrams. A review of yearly scram data from 1992 through 1996 revealed that manual scram rates have increased from 21 percent to 37.7 percent. The MRBI teams also found that most licensees count plant level MPFFs against SSCs that cause unplanned *automatic and manual* reactor scrams and that few, if any, licensees actually counted just unplanned *automatic* reactor scrams. Accordingly, the NRC added guidance in Revision 2 to RG 1.160, Regulatory Position 1.72, "Unplanned Manual Scrams," which states: "The staff's position is that all unanticipated scrams be considered, including those scrams that are manually initiated in anticipation of automatic scrams. The purpose of this is not to discourage manual trips but rather to ensure that operators do not mask a maintenance performance issue."

In addition, the MRBI teams found that some licensees monitored components failures (i.e, breakers, MOV actuators, solenoid-operated valves, limit switches, relays, etc) at the system or train level when functional failures of these components occurred across several different and diverse systems or system trains. In many instances, the systems would not exceed their performance criteria under paragraph (a)(2); however, a history of common-mode failures at the component level would necessitate the need to evaluate components for monitoring in accordance with paragraph (a)(1). See Section 2.5.2 of this report for additional details.

Conclusions

The MRBI teams concluded that licensees should monitor the RPS and ESFAS at the train (e.g., half scrams) and/or channel level. In addition, the MRBI teams also found it prudent that most licensees currently monitor both unplanned *automatic* and *manual* scrams.

Recommendations

The NRC staff recommends that all licensees should monitor RPS, ESFAS, and other HSS I&C systems at the train (e.g., half-scrams) and/or channel level. The NEI should also revise the plant level performance criteria in NUMARC 93-01 to include unplanned *manual* scrams.

2.5.6 Monitoring and Trending the Condition of Structures

The rule requires licensees to monitor the performance or condition of structures in a manner sufficient to provide reasonable assurance that those structures are capable of fulfilling their intended functions. The SOC for the rule states that where failures are likely to cause the loss of an intended function, monitoring should be predictive in nature, providing early warning of degradation before failures occur. NUMARC 93-01, Section 9.4.2.4, "Monitoring Structure Level Goals," and Section 10.2.3, "Monitoring the Condition of Structures," provide condition monitoring guidance for structures. RG 1.160, Regulatory Position 1.5, "Monitoring Structures," clarified the NUMARC guidance by providing a technical basis for determining when a structure is in a degraded condition; it should be moved to the paragraph (a)(1) monitoring category. In addition, the NRC used IP 62002, "Inspection of Structures, Passive Components, and Civil Engineering Features at Nuclear Power Plants," dated December 31, 1996, to verify that licensees established adequate condition monitoring programs for structures within the scope of the maintenance rule.

The MRBI teams reviewed licensees' programs for monitoring structures to ensure that they established appropriate performance or condition monitoring activities. These reviews emphasized that monitoring should be predictive in nature, providing early warning of degradation, and where practical, identify conditions where failures are likely to cause a loss of intended function. The MRBI teams also verified that the licensees established guidelines for moving structures from paragraph (a)(2) to paragraph (a)(1) when condition monitoring performance measures are exceeded or when a civil engineering analysis determines that a structure may fail before its next MR periodic inspection interval.

Findings

During the initial 18 MRBIs, some licensees assumed that most structures would be inherently reliable and, therefore, did not require monitoring under the maintenance rule despite the fact that there were already ongoing monitoring and preventive maintenance activities for structures at the site. Many of these structures are monitored during the normal course of operator and management inspection tours and observations by other plant departments in the course of their normal work activities. The MRBI teams identified 15 sites with structural monitoring program concerns because of inadequate NRC and industry guidance; therefore, the teams opened inspector followup items to reevaluate structural monitoring programs after proper guidance was developed. In addition, the MRBI teams found that some structures met the scoping criteria in 10 CFR 50.65(b), but were not included within the licensees' MR programs.

Following the initial 18 MRBIs, the NRC added Regulatory Position 1.5, "Monitoring Structures," to Revision 2 of RG 1.160. Specifically, the guidance presented additional condition monitoring methods for structures and established a technical basis for determining when structures should be moved to the paragraph (a)(1) monitoring category.

As previously noted on page 2-3 of this report, the NRC staff recognized the need for additional guidance on structural monitoring before the effective date of the rule. NUMARC 93-01, Revision 2, Section 10.2.3, "Monitoring the Condition of Structures," contains some additional guidance. However, the staff expected NEI to finalize comprehensive guidance developed in NEI 96-03, "Guideline for Monitoring the Condition of Structures at Nuclear Power Plants" (NEI, 1996). NEI 96-03 was intended to provide structural monitoring guidance for all regulatory applications, not just the maintenance rule. Although the NRC staff commented on NEI 96-03, NEI decided not to pursue NRC endorsement of the guidance.

After the NRC issued RG 1.160, Revision 2, licensees modified their structural monitoring programs to meet the guidance established in RG 1.160. The MRBI teams subsequently found that most licensees included the appropriate structures within the scope of the rule with a few exceptions. The teams also found that certain structures, such as the primary containment, were monitored using existing monitoring and testing requirements, including those in 10 CFR Part 50, Appendix J. Other structures, such as auxiliary, turbine, and switchyard buildings, were amenable to condition monitoring (e.g., monitoring water intrusion past seismic gaps, which caused MPFFs on components inside the building).

Some structures required engineering evaluations to establish condition monitoring criteria. Most licensees established structural monitoring programs, which included specific quantitative or qualitative criteria. Similarly, most licensees established predictive performance measures or goals providing early warning of significant degradation before failures occurred. The MRBI teams found that most licensees evaluated structures under paragraph (a)(2); however, some licensees found significant degradation, which resulted in a few structures being moved to the (a)(1) monitoring category with corrective actions being taken to schedule maintenance that would restore the design-basis condition of the structures in question.

Conclusions

During earlier MRBIs, the industry was not given adequate guidance on structural monitoring to meet the intent of the rule. Following the initial MRBIs, the NRC added Regulatory Position 1.5, "Monitoring Structures," to Revision 2 of RG 1.160. After receiving this additional guidance, almost all licensees developed structural monitoring programs that meet the monitoring requirements of the maintenance rule.

Recommendations

None

2.5.7 Monitoring Functional Failures Versus Maintenance Preventable Functional Failures

To meet the intent of the maintenance rule, the MRBI teams verified that licensees monitored MPFFs. In addition, the MRBI teams verified whether licensees monitored FFs, even though this is not required under the maintenance rule.

Findings

During earlier MRBIs, the inspection teams identified some FFs that were attributed to maintenance, but licensees did not always designate these failures as MPFFs. Not properly identifying MPFFs were cited as violations of paragraphs (a)(1) and/or (a)(2) if the licensee did not adequately monitor SSCs, if goals or performance measures were exceeded, or if preventive maintenance was ineffective. This problem in identifying MPFFs was due in part on licensees' perceptions that identifying MPFFs would reflect adversely on the effectiveness of their maintenance program.

Most licensees monitor both FFs and MPFFs under their MR monitoring program. In general, these programs define a FF as the failure of an SSC to perform its MR intended function. The MRBI teams also found that most licensees identified FFs that were sometimes attributed to maintenance, but in many cases the FFs were caused by other factors (e.g., operator error, design deficiency).

Conclusions

Licensees may choose to monitor FFs each cycle, rather than MPFFs each cycle, since random failures caused by FFs each cycle may give more complete information on equipment reliability. In addition, MPFFs each cycle should be considered a subset of FFs each cycle. Although licensees are encouraged to maintain conservatism in their monitoring programs, the MRBI teams noted that monitoring FFs each cycle is not required under the MR.

Recommendation

Licensees are encouraged to monitor both FFs and MPFFs, but this is not a requirement of the rule.

2.6 Periodic Evaluations in Accordance With Paragraph (a)(3)

The MRBI teams verified that the licensees generally completed a periodic evaluation once every refueling cycle, not to exceed 24 months as required by 10 CFR 50.65(a)(3). The purpose of these evaluations is to assess whether licensees need to adjust the goals, performance measures, and/or preventive maintenance programs for SSCs within the scope of the rule. These evaluations must consider both site-specific and industry-wide operating experience (IOE). Additionally, the rule requires licensees to make adjustments to ensure that the objective of preventing failures of SSCs through maintenance is appropriately balanced against the objective of minimizing unavailability resulting from monitoring or preventive maintenance.

Findings

The MRBI teams found that, in general, licensees completed the periodic evaluations within the prescribed time frames. However, a few licensees were late in completing the initial periodic evaluation. Most licensees used IOE information on reliability and availability performance of similar systems across the industry to make adjustments where necessary. Most licensees also used PRA methods to establish the technical basis for appropriate reliability and availability performance measures. If SSC performance measures were maintained within limits based on PRA assumptions and site-specific IOE data, the MRBI teams concluded that the licensee maintained a balance between reliability and availability. In addition, many licensees used the periodic evaluations to evaluate the overall effectiveness of their MR implementation programs and made adjustments where necessary to improve plant performance.

2.6.1 Use of Industry Operating Experience

The maintenance rule requires that, where practical, the periodic evaluations shall take into account IOE. This type of information was usually available in licensees' existing operating experience programs.

However, licensees didn't always revise their existing operating experience programs to address the requirements of 10 CFR 50.65(a)(1) and (a)(3).

The MRBI teams found that licensees' system engineers reviewed the IOE at regular time intervals (e.g., monthly) and incorporated lessons learned into the adjustment process to ensure an effective maintenance program. IOE sources included NRC bulletins, generic letters, and information notices, as well as vendor technical information letters (TILs) and General Electric (GE) service information letters (SILs), Westinghouse nuclear service advisory letters (NSALs) and nuclear service technical bulletins (NSTBs), as well as significant event reports (SERs) and significant operating experience reports (SOERs) promulgated by the Institute for Nuclear Power Operations (INPO). The plants evaluated this information and then incorporated it as lessons learned information that could change the preventive maintenance program or training program, where appropriate.

Licensees no longer use IOE from the INPO Nuclear Plant Reliability Data System (NPRDS) database because the database has been replaced by the Equipment Performance and Information Exchange (EPIX) System. The EPIX database contains INPO information and data collection commitments, which licensees download into their site-specific operating experience databases. Using this resource, licensees can compare reliability and availability data on similar SSCs within the scope of the rule at other plants. In addition, the EPIX database contains information on equipment reliability and availability data. This information can be used to adjust reliability and availability goals, performance measures, and preventive maintenance activities. This information can also be used to update actual industry-wide performance data and to update the reliability and availability assumption used in a plant-specific PRA.

2.6.2 Balancing Reliability Against Availability

The maintenance rule requires licensees to make adjustments where necessary to ensure that the objective of preventing failures of SSCs through maintenance is appropriately balanced against the objective of minimizing unavailability of SSCs caused by monitoring or preventive maintenance activities. The intent of this requirement is to ensure that monitoring or preventive maintenance activities do not cause excessive unavailability that would negate any improvement in reliability achieved as a result of the monitoring or maintenance activity, and that deferring monitoring or preventive maintenance to achieve high availability does not cause low reliability.

Because it is impractical to achieve balancing (or optimization) of reliability and availability on a continuous (day-to-day) basis, licensees should review their maintenance schedules periodically and make adjustments, where necessary, to improve preventive maintenance activities. Licensees must demonstrate that they have achieved a balance between reliability and availability during the periodic evaluation period (i.e., once every refueling cycle, not to exceed 24 months). Additional guidance is also presented in NUMARC 93-01, Section 12.2.4, "Optimizing Availability and Reliability for SSCs."

Findings

Some licensees had difficulty with the balancing process because appropriate reliability and availability performance criteria had not been established. In these cases, there was no clear technical bases for established performance measures. If inappropriate performance measures are used, a balance cannot be achieved.

Most licensees have implemented an appropriate method for evaluating maintenance activities and made adjustments where necessary every refueling cycle, not to exceed 24 months. During the early MRBIs, many licensees had not completed their initial 50.65(a)(3) periodic assessments; therefore, balancing was not evaluated. In some cases, no unavailability historical reviews were performed for HSS SSCs. Most licensees reviewed HSS SSCs that were subjected to this process and verified that adjustments were made to balance reliability and availability. This is an ongoing activity for licensees and the NRC, which may be used to measure the efficacy of the rule.

The MRBI teams found that most licensees achieved a balance between reliability and availability by verifying that SSCs met their performance criteria or goals. If the performance criteria or goals were not met, the licensees would initiate corrective actions needed to adjust the goals, performance measures, and/or maintenance activities so that a balance could be achieved.

At a few plants, the concept of condition probability (of success) was used as a performance measure to determine if a balance was being achieved between reliability and availability for HSS SSCs. The teams found that these licensees adequately used this method to evaluate the balance between reliability and availability. The conditional probability [of success] for operation of equipment was defined by the following equation:

$$CP = (P_A) * (P_S)_A * (P_R)_{A\&S}$$

where: P_A = Probability that equipment is available
        $(P_S)_A$ = Probability of successful starts given the equipment is available
        $(P_R)_{A\&S}$ = Probability of continuous runs given the equipment is available and starts successfully

The reduced form of this conditional probability [of success] equation defines a multiplication product of the availability and reliability variables. The preceding equation provides a means to calculate the maximum number of failures that HSS equipment may have without exceeding the conditional probability failure rate assuming 100 percent availability. In addition, the same methods can be used to determine the maximum allowable outage or unavailability time given perfect reliability. Thus, a proper maximum reliability and unavailability number can be derived from conditional probability, and licensees can set up tables showing failure rates and unavailability data for everything in between the derived conditional probability limits that a piece of equipment could experience during actual cycle operation and still meet its chosen conditional probability. The table failure rates and unavailability data limits could then be used as bounding limits for reliability and availability and compared to actual equipment performance to determine if a balance between reliability and availability is achieved, or if adjustments to equipment maintenance programs are warranted.

For HSS SSCs, monitoring plant specific reliability and availability data for specific SSCs (i.e., major pump, valves, emergency diesel generators, etc.) may be accomplished by establishing databases that can track and trend equipment performance to verify that a balance is being achieved between reliability and availability.

A concern with the use of a conditional probability measure for balancing reliability and availability was the potential masking of the reliability variable by the availability variable. For example, reliability could increase as a result of improvements in design or environmental conditions, and availability could decrease because of scheduling more preventive maintenance, which masks the effects of increasing reliability in the conditional probability formula. Licensees should be careful in their use of conditional probability alone as a performance measure since it can cause masking issues between reliability and availability. This is why most licensees established separate performance criteria for reliability and availability. Because the use of conditional probability alone as a performance measure for some HSS SSCs might not be sufficient, some licensees have discontinued the use of conditional probability as a measure of performance with respect to complying with the requirements of 10 CFR 50.65(a)(3)

Conclusions

The MRBI teams found that most licensees established appropriate methods for balancing reliability and availability. A few licensees used conditional probability as a PRA method to establish bounding limits between reliability and availability. These licensees tracked HSS SSC reliability and availability data separately and used the conditional probability value to statistically determine whether a balance was being achieved between reliability and availability. Licensees should be careful in their use of conditional probability alone as a performance measure since it can cause masking issues between reliability and availability. A few licensees also did not complete their periodic evaluations in a timely manner. Many licensees also used the periodic evaluations to determine the overall effectiveness of their MR implementation programs and made adjustments where necessary to improve plant performance.

Recommendations

Licensees may use the conditional probability method to determine the appropriate bounding limits for reliability and availability; however, they should ensure that performance measures for reliability and availability are established separately to avoid masking issues. If used properly, this method can verify that a balance is being achieved between reliability and availability to meet the requirements of 10 CFR 50.65(a)(3). Licensees can continue to use the periodic evaluations required by paragraph (a)(3) to determine the overall effectiveness of their MR implementation programs and make adjustments where necessary to improve plant performance.

2.7 Safety (Risk) Assessments Before Performing Maintenance

In accordance with paragraph (a)(3) of the maintenance rule, the NRC expects licensees to assess the impact on plant safety before taking plant equipment out of service for monitoring or preventive maintenance. This assessment is performed on an ongoing basis for all modes of operation (i.e., full power, transitional modes, and shutdown) and for all maintenance activities for SSCs within the scope of the rule.

As stated in the SOC dated July 10, 1991 (NRC, 1991), assessing the cumulative impact of out-of-service equipment on the performance of safety functions is intended to ensure that the plant is not placed in safety (or risk) significant configurations. These assessments do not necessarily require that a quantitative assessment of probabilistic risk be performed. However, the PRA or IPE may provide useful information on safety significance of various configurations. The level of sophistication with which such assessments are performed is expected to vary, based on these circumstances or prevailing conditions. The assessments may range from a simple matrix based on qualitative and quantitative risk insights to the use of an on-line living PRA tool or risk monitor. It is expected that, over time, assessments of this type will be refined as the technology improves and experience is gained.

The MRBI teams verified whether licensees adequately assessed the overall effect on the performance of safety functions when SSCs were removed from service for monitoring or maintenance. This included reviewing licensees' processes, procedures, and methods for taking equipment out of service (OOS) to perform maintenance. The focus of the reviews was to evaluate the results of safety assessments as an input to control the equipment outage configuration through a review of control room logs that track the status (in or out of service) of plant equipment. The risk estimates of selected configurations were quantitatively reviewed and the risk impacts of sampled configurations were qualitatively reviewed to ensure that undesirable configurations did not occur. Additional guidance is given in Section 11.0 of NUMARC 93-01.

Findings

The MRBI teams found that the nuclear industry has changed since the maintenance rule was issued in 1991. One significant change is that licensees have increased the frequency and amount of maintenance while at power. This may be caused by rate deregulation of the electric utility industry, which requires nuclear power plants to operate more efficiently. One mechanism for increasing efficiency is to shorten refueling outages and reduce or eliminate mid-cycle outages by performing more maintenance while at power. As discussed in a letter dated October 6, 1994, from the Director of the Office of Nuclear Reactor Regulation (NRR) to the Executive Vice President of the NEI, NRC senior management became concerned with both the increased frequency and amount of on-line maintenance and the apparent lack of licensees' understanding of the impact of maintenance on plant safety.

Following rule implementation after July 10, 1996, the MRBI teams found that licensees used a variety of approaches to assess the overall safety impact of taking plant equipment OOS for maintenance. The licensees' safety assessment processes were either quantitative, qualitative, or a combination of both. Where quantitative methods were used, the MRBIs verified that the PRA models used to quantify risk were of sufficient scope and quality to support the assessments. The same considerations used in evaluating the fidelity of the PRA models for determining safety significance were also applicable to the analytical models used for the safety assessments.

The specific format of the quantitative assessments used by licensees varied. However, the end result of these assessments provided information about the effects of individual maintenance configurations on plant risk. The specific measures of plant risk were usually clearly defined (e.g., CDF, LERF). In this respect, certain approaches have been shown to exhibit unique strengths and weaknesses, which are specific to the approach used. The assessments usually considered the risk impact associated with the proposed maintenance activities from SSCs used to mitigate events as well as the risk impact from SSCs that are considered to be event initiators (e.g., scheduling switchyard maintenance during an emergency diesel outage).

2.7.1 Risk-Informed Safety Assessment Procedures/Risk Monitors

A sophisticated method of performing safety assessments consisted of actually quantifying the proposed maintenance configurations using a full plant PRA model or "risk monitor." Tools of this type make it possible to analyze a wide variety of unique plant configurations. If this approach is used, the overall adequacy of the assessment depends on the underlying PRA model used in quantifying the configuration as well as the accuracy of the input assumptions regarding the availability of the equipment being considered for maintenance. Since PRA models have sometimes been simplified or optimized, the adequacy of the optimized model was reviewed to determine whether the PRA model accurately reflects the "as operated" plant baseline configuration. In particular, the scope of the PRA model may not cover SSCs related to containment performance. If SSCs affecting containment performance were not adequately modeled in the risk monitor tool, the output analysis would significantly underestimate total plant risk.

Some early versions of "risk monitors" used pre-solved cut sets. In such PRA models, very highly reliable SSCs may be truncated from the analysis. Thus, the fidelity of the results from these types of risk monitors decreases when truncated SSCs are out of service at the same time. If this type of risk monitor is used, then the truncation limits have to be carefully reviewed to determine its effect on the loss of result fidelity. Therefore, compensatory actions for maintenance configurations involving truncated SSCs also needed to be assessed.

Another approach involved using a matrix of pre-analyzed plant configurations. Typically, the risk matrix defines the unacceptable plant configurations for online maintenance. However, this method is limited by the number of pre-analyzed configurations that can be considered. For example, the matrix may not include some LSS SSCs because of size limitations. It is possible that some of these combinations of LSS SSCs may be out of service and their risk impact would not be fully evaluated. In particular, emergent conditions may result in SSC configurations that are not explicitly addressed in the risk matrix. Thus, the potential exists for a plant to be outside the scope of pre-analyzed conditions. As a minimum, the NRC expected licensees' programs to ensure that key plant safety functions are maintained whenever the maintenance configurations exceed the boundaries of pre-analyzed configurations.

Most licensees established and implemented an ongoing, documented process for assessing the overall safety impact of maintenance configurations before performing maintenance on SSCs within the scope of the rule. In addition, most licensees have developed databases to indicate which individual SSCs are in or out of service.

Currently, the NRC staff has not established guidance on acceptable thresholds for temporary increases in plant risk. As a result, most licensees evaluated during the MRBIs used the guidance contained in Topical Report (TR) 105396, "Probabilistic Safety Assessment (PSA) Applications Guide"; developed by the Electric Power Research Institute (EPRI, 1995). During MRBIs, when temporary risk changes caused by online maintenance activities were found to exceed the limits discussed in EPRI TR-105396, the MRBI teams evaluated whether the temporary changes in plant risk were reasonable and managed.

On August 1, 1997, the NRC staff issued SECY-97-173, "Potential Revision to 10 CFR 50.65(a)(3) of the Maintenance Rule to Require Licensees to Perform Safety Assessments." As discussed in SECY-97-173, where 21 site-specific safety assessment programs were evaluated, the MRBI teams found five sites at which safety assessments were not adequately performed. The licensees had programs for assessing risk; however, weaknesses in their programs or implementation processes contributed to not performing the assessments. At two sites, it appeared that the licensees simply did not perform the assessments in isolated cases. At one site, not performing the assessment resulted from emergent work occurring between the "freeze date" of the maintenance schedule and the start of the planned maintenance outage. At another site, additional maintenance activities were included after the schedule was pre-analyzed. At a third site, some of the equipment that was out of service was not assessed in the pre-analyzed risk matrix; therefore, the associated risk was significantly higher. The causes of not performing the assessments include inadequate procedures, insufficient involvement of PRA staff, not following procedures, and insufficient knowledge of the PRA insights by operators and scheduling staff.

Although the safety significance of the non-assessed maintenance configurations was not quantitatively determined during the MRBIs in all cases, it appears that some of the non-assessed configurations resulted in the plant being in a state of substantially greater risk than assumed. Further, the licensees' lack of awareness of the level of risk is of significant regulatory concern because weaknesses in the licensees' procedures that led to the lack of awareness could allow maintenance configurations of even greater risk to be entered without being adequately assessed. Given that the MRBI teams reviewed only a sample of the maintenance configurations, the staff considered the five missed assessments and their apparent risk significance to be a safety concern.

The other 11 sites had weaknesses in their paragraph (a)(3) safety assessment implementation ranging from minor training weaknesses to poor procedures and methods that did not include all HSS SSCs in the risk matrix. Most of these licensees adequately addressed the impact of removing HSS SSCs from service, but often did not have a clear method for assessing the removal from service of SSCs of LSS[10] or for assessing combinations outside of the pre-analyzed matrix. Another common finding involved procedural weaknesses for addressing emergent work. At these 11 sites, the inspectors did not identify specific instances of safety assessments not being performed in the sample of maintenance configurations reviewed; nonetheless, weaknesses in these programs raised the potential that an assessment may not have been performed.

In summary, the first 21 MRBI teams found that about 76 percent of the licensees inspected had weaknesses in their methods for performing paragraph (a)(3) safety assessments. In 24 percent of these inspections, the teams found instances where assessments were not performed. The staff believes that these findings do not meet the staff's or the Commission's current expectations regarding the recommendation in paragraph (a)(3).

After the NRC staff issued SECY-97-173, the MRBI teams completed the rest of the 68 inspections in which the teams reviewed the licensees' safety assessment programs conducted before performing

---

[10]Even though LSS SSCs were not on the matrix, the unavailability of LSS SSCs could, depending on other SSCs out of service, substantially increase risk.

maintenance. In general, the teams found that 26 sites (i.e., 38%) implemented good safety assessment programs with no or very minor weaknesses, 35 sites (i.e., 52%) implemented adequate safety assessment programs with some weaknesses, and 7 sites (i.e., 10%) did not perform adequate assessments before performing maintenance. Thus, of the 68 MRBIs, about 62 percent revealed some weaknesses with licensees' implementation of paragraph (a)(3) safety assessments, including instances of licensees not performing the assessments required by their programs. This evaluation is based on the conclusions reached by the NRC at the time of the MRBIs, and does not reflect licensees' current programs which have been updated to implement corrective actions to resolve NRC and licensee identified issues, weaknesses or upgrades that licensees initiated to improve safety assessment programs.

In addition, the MRBI teams found that some licensees did not perform safety assessments on all maintenance activities during all modes of operation. For example, one plant experienced a loss of the residual heat removal (RHR) system complicated by maintenance activities while shutdown. Other licensees did not consider surveillance test activities that could temporarily remove a system's automatic function and where complicated manual actions were needed to return the system to service.

Conclusions

The MRBI teams recognized that the nuclear power industry has changed since the maintenance rule was issued in 1991. One significant change is that licensees have increased the frequency and amount of maintenance while at power. One mechanism for increasing plant capacity factors is to shorten refueling outages and reduce or eliminate mid-cycle maintenance outages by performing more maintenance while at power. The NRC is concerned with both the increased frequency and amount of on-line maintenance and, in some cases, licensees' lack of understanding of its impact on plant safety. Therefore, the Commission directed the staff to revise 10 CFR 50.65(a)(3) to require licensees to perform safety assessments before performing maintenance.

Recommendations

The NRC staff should work with industry stakeholders to revise NUMARC 93-01, Section 11.0, "Evaluation of Systems To Be Removed From Service," which could be endorsed by RG 1.160, to address rulemaking activities on 10 CFR 50.65(a)(4). See Section 2.7.2 of this report for additional details. The NRC staff should work with the NEI, industry representatives, and stakeholders to revise RG 1.160 to produce adequate guidance for licensees to implement safety assessment programs that would comply with the requirements of 10 CFR 50.65(a)(4).

2.7.2   Rulemaking Activities on 10 CFR 50.65(a)(4), Performing Safety Assessments Before Removing Equipment From Service for Maintenance

On August 1, 1997, the NRC staff issued SECY-97-173, "Potential Revision to 10 CFR 50.65(a)(3) of the Maintenance Rule to Require Licensees to Perform Safety Assessments." The purpose of SECY-97-173 was to obtain Commission agreement with the staff's recommendation that the staff should propose rulemaking to revise the maintenance rule to require licensees to assess the impact on safety when removing equipment from service for maintenance. In SECY-97-173, the staff proposed three alternatives. The first alternative was not to change the rule. The second alternative would change the rule language from "should" to "shall" to require safety assessments before taking SSCs out of service for preventive maintenance. The third alternative would require a comprehensive revision to paragraph (a)(3). The staff would evaluate all three alternatives as part of the regulatory analysis. The final recommendation as to which alternative should be pursued would be based on the results of this regulatory analysis.

The staff concluded that it should proceed with the proposed rulemaking to require safety assessments before conducting maintenance for the following reasons: (1) the importance to safety of licensees' understanding of risk associated with maintenance configurations, (2) the increased performance of maintenance while at power, (3) the licensees' proposed use of their (a)(3) safety assessment programs in other risk-informed initiatives, and (4) staff experience gained with this issue during the MRBIs.

On December 17, 1997, the Commission approved the staff's recommendation to develop proposed rulemaking to revise the maintenance rule to require that safety assessments be taken into account before performing maintenance activities (NRC, 1997b), subject to the following comments:

(1)   Although all three alternatives, including not changing the rule, should be considered as part of the regulatory analysis for proposed rulemaking, extended or protracted regulatory analysis of Alternative 1 is unnecessary.

(2)   In addition to the change from "should" to "shall" in Section 50.65(a)(3) as proposed by the staff in Alternative 2, the proposed rule should also incorporate the following changes that are consistent with NRC Regulatory Guide 1.160, Revision 2, and NUMARC 93-01, Revision 2. The staff may suggest alternative wording for Commission consideration as part of the proposed rulemaking package, if the following rule language is problematic:

(a)   Since the requirements of the maintenance rule, including the assessment of SSCs proposed to be removed from service, are applicable during all modes of plant operation, the following clarification should be added as a preamble to the maintenance rule:

"The requirements of this section are applicable during all conditions of plant operation, including normal shutdown operations."

(b)   Revise the third sentence of (a)(3) to read as follows:

"Adjustments shall be made where necessary to ensure that the objective of *preventing* failures of structures, systems, and components through maintenance is appropriately balanced against the objective of minimizing unavailability of structures, systems, and components due to monitoring or *preventive* maintenance."

(c)   The final sentence of Section 50.65 (a)(3) should be designated as (a)(4) and revised as follows:

"Before performing maintenance activities on SSCs within the scope of this section (including, but not limited to, surveillance testing, post-maintenance testing, corrective maintenance, performance/condition monitoring, and preventive maintenance), an assessment of the current plant configuration as well as expected changes to plant configuration that will result from the proposed maintenance activities shall be conducted to determine the overall effect on performance of safety functions. The results of this assessment shall be used to ensure that the plant is not placed in risk-significant configurations."

(3)   Since the changes to the maintenance rule are part of a larger set of initiatives, including, but not limited to, changes to 10 CFR 50.59 and the integrated review of the NRC assessment process for commercial reactors, the staff should ensure consistency among these efforts.

(4)   In the limited regulatory analysis discussion of Alternative 3, staff should briefly consider how this alternative might be pursued. One disadvantage of Alternative 2 is that licensees could theoretically use technically inferior methods for conducting safety assessments and could theoretically perform maintenance in configurations involving risk levels that may be imprudent, yet still argue that they are in compliance with the requirements of the revised maintenance rule to take into account safety assessments before performing maintenance.

To address this issue, the Commission would consider, as part of a future separate rulemaking, a staff proposal to incorporate by reference updates to NUMARC 93-01, Revision 2, and NRC Regulatory Guide 1.160, Revision 2, which emerge from the activities described in item 5.

(5)   As part of the regulatory guidance for this proposed rulemaking, the staff should supplement and expand on the discussion that was presented in the SOC for the original maintenance rule with regard to (a) variations in the rigor and sophistication of the assessments depending on the number and safety significance of SSCs out of service and (b) NRC's general expectations with regard to risk levels that the assessment should take into account to ensure a plant is not placed in risk-significant configurations during maintenance activities. This discussion should acknowledge that there are several assessment tools used to determine the risk significance of plant configurations, including PRA, deterministic analysis, considerations of defense in depth, and qualitative measures. This discussion would be short of the comprehensive treatment in Alternative 3 of these issues in the rule itself and would not constitute binding regulatory requirements. In developing this guidance, the staff should also consider whether the "Guidelines for Industry Actions to Assess Shutdown Management," NUMARC 91-06, as referenced in Section 11.2 of NUMARC 93-01, Revision 2, could be endorsed by the NRC. Consistent with the Commission's decision on Direction Setting Issue (DSI) 13, "The Role of the Industry," the staff should work with stakeholders in developing regulatory guidance. Development of regulatory guidance should not delay NRC staff efforts to issue the proposed rule.

In its SRM on SECY-97-173 (NRC, 1997b), the Commission also indicated that development of the regulatory guidance should not delay issuance of the proposed maintenance rule change. The staff initially planned to prepare a regulatory guide in conjunction with the development of the final rule and have it ready for issuance 120 days after the publication date of the rule. The Nuclear Energy Institute (NEI), representing the nuclear power industry specifically recommended in a letter dated October 10,

1997, to the NRC Executive Director for Operations, that the word "should" in the last sentence in paragraph (a)(3) be changed to "shall" and should be "made immediately effective" (NEI, 1997b). At this time, NEI was developing changes to NUMARC 93-01 to implement the revised rule.

On July 2, 1998, the staff issued SECY-98-165, "Proposed Revision to 10 CFR 50.65(a)(3) to Require Licensees to Perform Safety Assessments." The purpose of SECY-98-165 was to obtain the Commission's approval to publish a proposed rule requiring safety assessments before removing equipment from service for maintenance. Following the Commission's direction in the SRM on SECY-97-173, this proposed rule would

(1)   Add an introductory sentence to 10 CFR 50.65 clarifying that the rule applies under all conditions of operation, including normal shutdown. The rule applies during all operating conditions including shutdown, but it did not state this explicitly. As a preamble to 10 CFR 50.65, before paragraph (a)(1), the following statement would be added: "The requirements of this section are applicable during all conditions of plant operation, including normal shutdown operations."

(2)   Be silent regarding the change to the third sentence in paragraph (a)(3) because the desired editorial corrections were incorporated in the 1998 edition of the *Code of Federal Regulations*. The first "preventative" was corrected to "preventing," and the second "preventative" was changed to "preventive" for consistency.

(3)   Delete the last sentence of paragraph (a)(3) of the rule. This action removes and separates the safety assessment requirement from the more programmatic periodic evaluation requirements of the rule left in paragraph (a)(3).

(4)   Add a new paragraph (a)(4) that requires the performance of safety assessments. The industry guidance document, "Industry Guideline for Monitoring the Effectiveness of Maintenance at Nuclear Power Plants," NUMARC 93-01, includes the safety assessments in the recommended program. The NRC endorsed that guidance in RG 1.160. The industry's practice has been to incorporate the provision for performing safety assessments into its maintenance rule implementation programs, and NRC's intent has been that the safety assessments be performed. However, at present, the NRC cannot ensure that licensees follow their own programs with regard to the safety assessments.

The new paragraph (a)(4) would read as follows:

> Before performing maintenance activities on structures, systems, or components within the scope of this section (including, but not limited to, surveillance testing, post-maintenance testing, corrective maintenance, performance/condition monitoring, and preventive maintenance), an assessment of the current plant configuration as well as expected changes to plant configuration that will result from the proposed maintenance activities shall be conducted to determine the overall effect on performance of safety functions. The results of this assessment shall be used to ensure that the plant is not placed in risk-significant

configurations or configurations that would degrade the performance of safety functions to an unacceptable level.

(5)   Specify that the scope of the requirement for performing those assessments is to cover all planned maintenance activities. The NRC's intent is that the performance of safety assessments not be limited to monitoring and preventive maintenance activities. It would be imprudent of the NRC to require the development of a licensee practice for reviewing the safety aspects of most maintenance activities while omitting safety assessments before planned corrective maintenance activities. In fact, many licensees have already voluntarily included in their programs the performance of safety assessments before all planned maintenance activities.

(6)   Specify that the safety assessments are to examine the existing plant condition and the condition expected during the maintenance activity. The proposed language would be more specific regarding the reviews of (a) actual plant conditions before the planned maintenance activity is begun and (b) plant conditions expected while the activity is in progress. The rule would require the licensee to recognize its plant's capability to perform safety functions both before and after the planned change in plant condition for the performance of the maintenance activity.

(7)   Specify that the results of the safety assessments are to be used to help the licensee ensure that the plant is not placed in risk-significant configurations, that is, a configuration for which the incremental contribution to the annual risk is not insignificant or configurations that would degrade safety functions to an unacceptable level. The proposed language would be more specific regarding the purpose of the safety assessments.

After the NRC staff issued SECY-98-165, the Quality Assurance, Vendor Inspection, Maintenance and Allegations Branch (IQMB) within the Office of Nuclear Reactor Regulation (NRR), held public meetings with NEI, the Advisory Committee on Reactor Safeguards (ACRS), and the Commission to discuss the final rule language in 10 CFR 50.65(a)(4). The NRC proposed changing the rule language to that proposed in SECY-97-173 as follows:

> Before performing maintenance activities (including but not limited to surveillance, post-maintenance testing, and corrective and preventive maintenance), the licensee shall assess and manage the increase in risk that may result from the proposed maintenance activities.

The NRC staff received public comments indicating that the scope of SSCs under the maintenance rule safety assessment program was too broad. The NEI and industry suggested that licensees should be given flexibility in their safety assessment programs to limit the scope of 10 CFR 50.65(a)(4) to safety-significant SSCs as determined by a risk-informed evaluation process. See Section 2.11.1, "NRC Public Meetings With Industry Stakeholders During the Inspection Process," for additional details on this issue. After considering these public comments, the staff proposed the following final rule language:

> Before performing maintenance activities (including but not limited to surveillance, post-maintenance testing, and corrective and preventive maintenance), the licensee shall assess and manage the increase in risk that may result from the proposed maintenance

activities. The scope of the assessment may be limited to structures, systems, and components that a risk-informed evaluation process has shown to be significant to public health and safety.

On April 8, 1999, the NRC staff met with the ACRS to discuss draft regulatory guide (DG) 1082, "Assessing and Managing Risk Before Maintenance Activities at Nuclear Power Plants." At this public meeting, the NRC staff presented acceptable methods in DG-1082 for licensees to implement safety assessment programs before removing SSCs from service for maintenance. DG-1082 information will be incorporated into Revision 3 to RG 1.160. Details on comments from the public and the industry on the proposed rule, the final rule, and the regulatory analysis are presented in SECY-99-133, "Final Revision to 10 CFR 50.65 to Require Licensees to Perform Safety Assessments Before Performing Maintenance," dated May 17, 1999. Additional details on the development of these documents are given in Section 2.11 of this report. As discussed in Section 2.12 of this report, all of this information is also available on the NRC Maintenance Rule Home Page on the World Wide Web.

On June 18, 1999, the Commission issued an SRM approving the final rule language noted in the staff recommendations on SECY-99-133. The Commission directed the NRC staff to amend 10 CFR 50.65 to require that, before performing maintenance activities, power reactor licensees shall assess and manage the increase in risk that may result from performing maintenance activities. The Commission also directed the staff to complete RG 1.160 for NRC licensees and inspectors that will not become effective until 120 days after RG 1.160, Revision 3, is completed and in the public domain. Before issuing the final regulatory guide, the staff should send it to the Commission for review and approval.

Conclusion

The NRC staff concludes that 10 CFR 50.65(a)(4) shall be revised in accordance with the staff's recommendations in SECY-99-133 and the Commission approved SRM on SECY-99-133.

Recommendations

The NRC staff should continue to work with the NEI and industry stakeholders to revise NUMARC 93-01, Section 11.0, "Evaluation Of Systems To Be Removed From Service." The staff should revise RG 1.160, Revision 3, in efforts that can lead to endorsement of applicable changes to NUMARC 93-01, and implementation of rulemaking requirements on 10 CFR 50.65(a)(4).

2.8 Quality Assurance Audits/Self-Assessments

The NRC encourages licensees to perform self-assessments to identify potential violations and needed improvements in licensees' MR implementation programs. The NRC also gives licensees credit for identification, root cause analysis or cause determination, and corrective action for violations of the MR in its enforcement policy. These actions may be used to mitigate the violation to a non-cited violation, which does not have to be addressed in the inspection report. This credit is given if the licensee addresses the issue in its corrective action program and actions are taken to resolve the issue in a timely manner. However, if licensees identify a maintenance rule violation through a self-assessment but do not complete corrective actions, the utility could be cited for the violation. For information on NRC enforcement policy, individuals may access the *enforcement process* on the NRC Office of Enforcement's Home Page Web site at **http://www.nrc.gov/OE/process/htm**.

Findings

While conducting MRBIs, the teams reviewed licensees' quality assurance (QA) audits or self-assessments of MR programs, procedures, and implementation practices. The teams found that licensees' self-assessments generally identified several weaknesses in MR programs and implementation practices. The self-assessments typically identified problems with licensee activities regarding MR scope, performance measures and goals, tracking unavailability and MPFFs, timely completion of periodic assessments, making adjustments to maintenance programs, and safety assessment programs before taking SSCs out of service for maintenance. These issues were normally identified by licensees' QA organizations before MRBIs were conducted at each site. In most cases, the licensees identified the issues and took prompt corrective actions to address MR program and implementation problems. However, in a few cases, licensees did not complete changes to correct identified problems before the MRBI team arrived on site; therefore, the NRC could not give these licensees credit for implementing corrective actions in accordance with NRC enforcement policy.

Conclusions

Most licensees' self-assessments were effective at identifying weaknesses and areas needing improvements in MR programs. The NRC regarded the self-assessments as a MR program implementation strength.

Recommendations

Licensees should continue self-assessment activities periodically to verify that MR programs do not lapse and continue to demonstrate the effectiveness of maintenance activities in maintaining and improving overall plant performance.

2.9 Maintenance Rule Inspections at Plants in a Decommissioning Status

On August 28, 1996, the NRC amended the maintenance rule, specifically 10 CFR 50.65(a)(1), to include plants in a decommissioning status under 10 CFR 50.82. As amended, paragraph (a)(1) of the rule requires licensees to monitor the performance or condition of all SSCs associated with the storage, control, and maintenance of spent fuel in a safe condition in a manner sufficient to provide reasonable assurance that such SSCs are capable of performing their intended functions. After the rule was amended, Regulatory Guide 1.160 was revised (Revision 2, dated March 1997) to address the amended provisions for plants in a decommissioning status.

The NEI subsequently submitted an industry white paper, "10 CFR 50.65 Maintenance Rule Implementation for Shutdown Plants," dated August 26, 1997 (NEI, 1997a), seeking NRC approval for this guidance. In the white paper, NEI stated, in part, that "statements of paragraph (b)(1) and (b)(2) of the rule are not applicable to decommissioned plants." The NRC agrees that the scope of the MR for decommissioned plants is essentially described by paragraph (a)(1), but there is a subset of SSCs described in 10 CFR 50.65(b) that perform functions that apply to the storage, control, and maintenance of spent fuel in a safe condition (i.e., the spent fuel pool structure must remain functional following design-basis events such as earthquakes, tornadoes, etc.). Therefore, the NRC did not endorse this document. The NRC encouraged the NEI and the industry to formulate an acceptable guidance document on this subject, focusing on those elements that are necessary to comply with 10 CFR 50.65 rather than emphasizing those aspects which need not be complied with. To date, the NEI has not submitted an updated revision to the industry white paper to address this issue.

The NRC also held a workshop at the NRC Region III office on April 30, 1998, to comment on needed revisions to the industry white paper. Other principal topics that were discussed at the workshop included scoping issues, risk ranking practices, categorization of SSCs, periodic assessments, balancing of reliability and availability, and the level of performance or condition monitoring for all SSCs associated with the storage, control, and maintenance of spent fuel in a safe condition.

As presented at the workshop, the staff's position is that plants in a decommissioning status need to focus on design functions necessary for preserving the spent fuel in a safe condition in order to identify the relevant SSCs within the scope of paragraph (a)(1) of the rule. The staff emphasized that although the population of SSCs within the scope of the rule for plants in a decommissioning status would be significantly smaller than for operating plants, the implementation of the rule would remain fundamentally the same. The NRC staff also gave the industry a conceptual listing of SSCs that should be considered in licensees' MR implementation programs for plants in a decommissioning status. See Appendix D of this report, "Examples of SSCs That May Be in the Scope of the Maintenance Rule for Plants in a Decommissioning Status."

As discussed at the workshop, the effectiveness of the MR implementation process would continue to be evaluated by licensees of plants in a decommissioning status with a frequency not to exceed 24 months between evaluations. In particular, the current 24-month evaluation period for plants in a decommissioning status began at the time the rule was amended in August 1996. Therefore, the staff anticipates that licensees of plants in a decommissioning status will complete their paragraph (a)(3) periodic evaluations by August 1998. The staff also indicated that licensees would be expected to balance reliability and availability of SSCs that are within the scope of the rule for plants in a decommissioning status as a function of the periodic assessments.

During the workshop, representatives of Commonwealth Edison presented a brief discussion related to the development of the maintenance rule program for decommissioning at the Zion Station. The presentation was informative and afforded the workshop participants an opportunity to understand Commonwealth Edison's methodology for implementing the rule at a plant in a decommissioning status.

The NRC-sponsored workshop for implementation of the rule at plants in a decommissioning status was beneficial in identifying various performance-based approaches for implementing the rule. In addition, the NRC completed MRBIs at three plants in a 10 CFR 50.82 decommissioning status. The overall results are presented below.

Findings

As of the date of this report, the NRC has only conducted MR inspections at three plants in a decommissioning status: Zion, Big Rock Point, and LaCrosse BWR. The individual plant findings follow.

Zion Nuclear Plant

At the Zion Station, the NRC conducted a MRBI on June 15-19, 1998. This inspection identified the following findings concerning the licensee's MR program for a plant that entered into a decommissioning

status in May 1998:

The licensee's process for identifying SSCs within the scope of the MR was good. The licensee identified all SSCs and SSC functions for a shutdown plant that were within the scope of the MR program. The expert panel effectively identified important system functions, established acceptable performance measures and goals, conservatively assessed functional failures and appropriately categorized SSCs for monitoring or tracking under paragraphs (a)(1) or (a)(2). The structural monitoring program, as designed, was also acceptable; however, at the time of the MRBI, the structural monitoring baseline inspections were not completed. The licensee's classification of structures in accordance with paragraph (a)(1) was conservative and appropriate. Guidance for conducting periodic assessments was good, but was being revised to include a new approach for balancing reliability and availability for the shutdown plant. The process for assessing risk to spent fuel resulting from equipment being taken out of service was determined to be acceptable. System engineers with maintenance rule responsibilities were knowledgeable of MR requirements as well as industry operating experience requirements and practices. System engineers were also experienced and knowledgeable about their systems. Corrective actions identified by the licensee's self-assessments were effective in establishing a program that complied with MR requirements.

Big Rock Point Station

At the Big Rock Point Station, the NRC conducted a MRBI from May 4-8, 1998. The NRC identified the following issues concerning the licensee's MR program for a plant placed into a decommissioning status on August 29, 1997.

The scope of the Big Rock Point MR program included systems and functions related to the control, storage, and maintenance of spent fuel in a safe condition as required by paragraph (a)(1) of the rule. These systems and functions involved electrical power, spent fuel pool integrity, spent fuel and load handling, heat removal, water level, and water chemistry. No SSCs or functions were identified that were improperly evaluated as not being within the scope of the rule.

However, the team determined that the licensee did not adequately monitor SSCs in accordance with the requirements of paragraph (a)(1) nor did it adequately demonstrate effective preventive maintenance for SSCs tracked by paragraph (a)(2). Two Level IV violations were identified: (1) The licensee did not adequately demonstrate that the performance or condition of functions of the reactor cooling water, station power, spent fuel storage rack, and fuel bundle configuration systems had been effectively controlled by performing appropriate preventive maintenance in accordance with the requirements of 10 CFR 50.65(a)(2). Specifically, the licensee established inappropriate plant level performance criteria that would not demonstrate effective preventive maintenance, which would ensure that these systems remained capable of performing their intended function as required; and (2) the licensee did not adequately demonstrate that the performance or condition of functions of the demineralized water, emergency power, fuel handling, and radiation monitoring systems had been effectively controlled by performing appropriate preventive maintenance in accordance with the requirements of 10 CFR 50.65(a)(2). In this instance, the measures used in the demonstration for these system functions consisted only of an inappropriate reliability measure that allowed repetitive MPFFs that are a measure of corrective action rather than reliability. As a result, the licensee did not demonstrate that effective preventive maintenance ensured that the functions of these systems remained capable of performing as required.

The risk ranking and establishment of performance criteria for SSC functions done by the expert panel during the inspection were appropriate. Since the plant did not have a PRA for decommissioning, evaluations of risk ranking were based on plant experience and engineering judgment, which included review of function importance, probability of failure occurrence, and consequences of failure. Supporting documents used in the evaluations were SL-5203, "Zircaloy Oxidation Analysis," the updated final hazards summary report for decommissioning, and the decommissioning plan. Given the status of the plant, all functions within the scope of the maintenance rule were determined to have low safety significance. The NRC found this to be acceptable.

No conclusions could be drawn regarding paragraph (a)(3) periodic evaluations because the licensee's procedures were being revised to more appropriately address periodic evaluations for a decommissioned plant.

Since the plant was permanently shutdown and defueled, the licensee maintained a defense-in-depth strategy for key spent fuel pool safety functions on the plan of the day. The four key functions were decay heat removal, spent fuel pool makeup, ac power sources, and radiation monitors. The first three functions had primary and backup systems identified. A system could be removed from service using an approved contingency plan. Given the status of the spent fuel and the defense-in-depth strategy, the licensee was adequately assessing the risk of taking equipment OOS for maintenance. The processes for assessing plant risk resulting from equipment being OOS for shutdown risk management were also determined to be adequate.

The NRC examined the goals and corrective actions for the diesel fire pump, which remained in a paragraph (a)(1) status for the de-fueled plant. The pump was placed into the (a)(1) monitoring category because of three occurrences of a bulging hose in the jacket water heating system. Although the hose did not fail and no functional failures occurred, plant management chose to put the diesel fire pump in category (a)(1) to focus on corrective actions. The licensee selected a goal of no bulging hoses for 1 year. This was considered appropriate, and the NRC found that this condition monitoring goal was acceptable.

The initial structural monitoring baseline inspection was well done. For example, the expert panel corrected a weakness when licensee personnel did not establish periodic followup inspections.

Additionally, the MRBI team determined that IOE was appropriately available and considered for the development of performance criteria, goals, and corrective actions to restore performance.

The licensee's self-assessment clearly identified problems with the MR implementation, yielded a set of major issues, and gave the licensee an opportunity to begin an aggressive improvement program. During the inspection, the MRBI team determined that the licensee had not completed the actions defined in the response to the self-assessment.

LaCrosse Boiling-Water Reactor (BWR)

At the LaCrosse–BWR Station, the NRC conducted a MRBI on May 18–22, 1998. The NRC found the following issues for a plant placed into decommissioning status following shutdown on April 30, 1987.

The NRC concluded that the MR had not been properly implemented at the LaCrosse BWR power plant. Although the NRC concluded that the spent fuel was being maintained in a safe condition, the elements required by the MR were not effectively implemented into the plant's existing program. Two apparent violations were identified: (1) the licensee did not identify SSCs required to be monitored under the MR, and existing lists of maintained systems were inappropriate for MR applications; and (2) goals and monitoring were not adequately implemented for SSCs required to be monitored in accordance with the MR. The licensee's preventive maintenance program and surveillance program provided adequate goals and monitoring for some but not all of the required systems. The licensee's corrective actions were appropriately implemented.

The licensee concluded that the existing programs and procedures provided reasonable assurance that SSCs were monitored as required by the maintenance rule; however, the team was not able to confirm this conclusion during the inspection.

The MRBI team determined that guidance was not developed to perform the 24-month periodic assessments, including the methodology for balancing reliability and unavailability. The team also concluded that given the status of the spent fuel and a work planning strategy, the licensee was adequately assessing taking equipment out of service for maintenance.

The MRBI team also determined that the plant staff understood that applicable surveillance tests and preventive maintenance were being performed on the systems required to maintain the spent fuel in a safe condition; however, the licensee did not clearly understand the application of the 10 CFR 50.65(a)(1) monitoring program for a plant in a decommissioning status.

On the basis of the MRBI results, an NRC regional inspector conducted a followup inspection in the spring of 1999 to evaluate the licensee's corrective actions to include required systems within the scope of 10 CFR 50.65(a)(1) for a plant in a decommissioning status and to implement adequate monitoring programs for systems needed to support the functions of storing the spent fuel in a safe condition. The inspector verified that corrective actions were taken to include the appropriate systems within scope with goals or performance measures established and monitoring programs implemented. The inspector closed both open items.

Conclusions

With the exception of one plant (LaCrosse BWR), the NRC concluded that licensees developed successful methods for implementing the MR for plants in a decommissioning status. Although the population of SSCs within the scope of the MR is much smaller for plants in a decommissioning status, many of the methods used to monitor the performance or condition of SSCs needed to maintain the spent fuel in a safe condition are similar to the methods described in NUMARC 93-01 (i.e., monitoring reliability, availability, or condition) for an operating plant. For additional inspection report details on MRBIs at plants in a decommissioning status, refer to Section 2.12, "NRC Maintenance Rule Home Page on the Work Wide Web."

Recommendation

The NRC should work with the NEI and industry stakeholders to enhance industry guidance that specifies standard methods for implementing the MR at plants in a decommissioning status.

2.10   Maintenance Rule Enforcement Actions

The NRC used a MR enforcement panel to provide for consistent inspection and enforcement of the rule while conducting MRBIs. The enforcement panel consisted of the responsible regional branch chief, the HQMB branch chief and section chief responsible for oversight of the MR, the MRBI team leader, and an enforcement specialist from the Office of Enforcement (OE). For more information on NRC enforcement, individuals may access the NRC Office of Enforcement's Home Page Web site at **http://www.nrc.gov/OE**.

The overall results of the MRBIs indicated that most licensees did a good job of implementing a maintenance rule program that generally complied with the requirements of the rule. Licensees identified almost all of the SSCs that needed to be included within the scope of 10 CFR 50.65(b). Most licensees generally complied with the requirements of 10 CFR 50.65(a)(1), especially when licensees focused their attention on problem SSCs needing improved performance. On several occasions, licensees did not establish adequate goals for monitoring SSCs in accordance with 10 CFR 50.65(a)(1) or performance measures that would demonstrate effective preventive maintenance in accordance with 10 CFR 50.65(a)(2). Most licensees completed their periodic evaluations in a timely manner to comply with 10 CFR 50.65(a)(3) and developed assessment programs before taking equipment OOS for maintenance to satisfy the recommendation to perform assessments in 10 CFR 50.65(a)(3); however, the NRC found several weaknesses in licensees' pre-maintenance safety assessment programs.

A summary of the 68 MRBIs revealed the following findings concerning enforcement actions. Four plants received Level III violations with civil penalties because of a programmatic breakdown of the licensees' MR implementation efforts. One plant received a Level III violation with no civil penalty. This licensee's MR program contained programmatic breakdown elements for not monitoring the effectiveness of maintenance, but the licensee's prompt corrective actions and overall plant equipment performance results were mitigating factors justifying no civil penalty. Another plant's MR implementation efforts constituted a programmatic breakdown; however, the plant had been shutdown for 18 months at the time of the inspection, and the NRC issued a confirmatory action letter to require the licensee to rectify all its major problems, including MR issues, before restart. Therefore, the plant received enforcement discretion. In addition, MRBI teams found 23 Level IV violations of 10 CFR 50.65(a)(1), 51 Level IV violations of 10 CFR 50.65(a)(2), 10 Level IV violations of 10 CFR 50.65(a)(3), and 30 Level IV violations of 10 CFR 50.65(b).[11] The MRBI teams also identified some non-cited violations that were either already corrected by licensees or were isolated administrative errors with no safety significance.

The NRC conducted three MRBIs at plants in a decommissioning status. One plant did not meet any of the requirements in 10 CFR 50.65(a)(1). This plant was given enforcement discretion since the safety significance of the situation was considered minor; however, the regulatory significance of not complying with any of the aspects of this regulation was substantial. Another plant did not demonstrate that the performance or condition of certain SSCs that maintain the spent fuel in a safe condition were effectively controlled through the performance of appropriate preventive maintenance to meet the requirements of 10 CFR 50.65(a)(2). The third plant had a well established MR program that met all aspects of 10 CFR 50.65.

Conclusions

The NRC concluded that the enforcement panel was an effective method of maintaining inspection and enforcement consistency for all MRBIs. In addition, each MRBI team employed a staff support member (SSM) from HQMB to help maintain inspection and enforcement consistency.

Recommendations

The NRC should continue to use the MR enforcement panel to maintain inspection and enforcement consistency for all MR issues identified by regional inspectors. The NRC staff should also revise the enforcement guidance memorandum (EGM) on MR enforcement to incorporate lessons learned from MRBI enforcement actions.

2.11 NRC Public Meetings With Industry Stakeholders During the Inspection Process

While conducting MRBIs, the NRC held five public meetings with the NEI, Winston & Strawn, industry representatives, and industry stakeholders to discuss various aspects of MR implementation. Topics discussed at these meetings included the adequacy of MR implementation guidance, methods used to implement the first risk-informed performance-based rule, the NRC staff's initial focus on program-oriented versus results-oriented aspects of the rule, specific issues associated with MR implementation practices, and future plans for monitoring the efficacy of the rule. In addition, HQMB developed a Maintenance Rule Home Page on the World Wide Web to enhance communications between the industry, industry stakeholders, and the NRC concerning MR implementation issues. See Section 2.12 for additional details. The results of NRC communications with industry stakeholders are given below.

---

[11]This enforcement count is based on violations counted in each inspection report. The violations were not counted by plant docket number. In many of the violations noted above, the NRC internal MRBI open item database counts the violation as two violations (i.e., one for each plant docket number at a tw- unit site).

LESSONS LEARNED FROM MRBIs

Following the first 18 MRBIs, the NRC staff held a public meeting with NEI, licensees, NSSS vendor representatives, and the public on January 9, 1997, to reach a "common understanding" on the staff's positions regarding the following MR inspection issues:

- Technical basis for using MPFFs
- Use of questions and answers (Q&As) from the 1993 MR workshop
- Scoping
- Timeliness of dispositioning SSCs from (a)(2) to (a)(1)
- Technical guidance for monitoring structures
- Perceived prescriptiveness in MR inspections

The specifics included the NRC's and licensees' understanding of the following:

(1)   Use MPFFs as a reliability indicator that can be appropriately linked to the reliability assumptions in the PRA for HSS and LSS standby SSCs;

(2)   Clarify the 1993 Q&A workshop issues on operator errors that can cause MPFFs when mistakes are made returning SSCs to service following maintenance;

(3)   Include emergency lighting and communications within the scope of the MR since they perform significant accident mitigating support functions in executing the EOPs;

(4)   Perform timely cause determination and corrective actions when moving SSCs from (a)(2) to (a)(1) but flexibility on timeliness is given based on the safety significance of the issue;

(5)   Include structures in the (a)(1) monitoring category when significant degradation indicates that they may not perform their design-basis by the next periodic inspection; and

(6)   Focus on more performance-oriented aspects of the rule once the MR program-oriented baseline inspections are complete.

Additional details on these issues were given in an NRC memorandum from Suzanne C. Black, Chief, HQMB, to David B. Matthews, Chief, Generic Issues and Environmental Projects Branch (PGEB), dated January 17, 1997 (NRC, 1997a).

On April 23, 1998, members of the NRC staff held a public meeting with representatives from Winston & Strawn Maintenance Rule Inspection Clearinghouse (MRIC), the NEI, licensees, NSSS vendors, and the public on MR related issues. The meeting was held at the request of Winston & Strawn representing their Maintenance Rule Inspection Clearinghouse (MRIC), to discuss MR issues and the industry's general observations regarding the results of the MRBIs to date.

In general, the industry understood that the baseline inspections were intended to be programmatic in nature; however, it expressed hope that once the baseline inspections were complete, there would be a further transition to a performance-based regulatory philosophy as envisioned by the Commission. Specific industry observations included the following:

- The number of violations was higher than the industry had anticipated. Further training and NRC headquarters oversight by the enforcement panel and HQMB should be maintained to promote inspection consistency.

- The NRC should continue open dialogue and communications with the industry to improve licensees' understanding of NRC expectations with regard to MR implementation.

- The NRC enforcement actions for the Fort Calhoun steam line rupture appeared to be inconsistent with SECY-97-055, where from a regulatory standpoint, the occurrence of an MPFF is not a violation of the MR.
  MR implementation is very resource intensive. Most licensees used two to three full-time employees to implement the MR program and developed sophisticated databases needed to maintain real-time performance trending, monitoring, and reporting. The industry encourages the NRC to allow cost-effective approaches to implementing the rule. The NRC should also allow MR program changes that reduce or eliminate overly burdensome process steps.

- Overall, the MR appears to be contributing to better plant performance. The improvements appear to be mainly due to licensees' enhancements related to implementation of improved maintenance and operations programs. The data on unplanned scrams and safety system actuations continues to show a downward trend while capacity factors continue to increase.

The staff's response to these observations follows:

- The NRC staff provided an SSM from the HQMB on every MRBI team to promote inspection consistency throughout the industry. HQMB also continues to train regional personnel on MR implementation programs and processes. In addition, the enforcement panel will remain in place to promote enforcement consistency between regions.

- The NRC developed a Maintenance Rule Web Page to help promote communication with the industry on NRC staff expectations with respect to MR implementation programs and practices. Refer to Section 2.12 for additional details on this issue.

- With respect to the Fort Calhoun enforcement action, the SOC presented the NRC's position related to erosion/corrosion (E/C) monitoring programs. Specifically, the SOC states that "where failures are likely to cause loss of intended function, monitoring should be predictive in nature, providing early warning of degradation. Monitoring can be performance-oriented or condition-oriented (parameter trending) or both." In addition, licensees should use existing monitoring programs (e.g., E/C monitoring) to implement the monitoring requirements of the rule. In the case of E/C monitoring, the NRC noted significant weaknesses in this monitoring program that could have precluded the steam line rupture event at Fort Calhoun if the licensee had implemented the E/C program properly. While the NRC agrees that the occurrence of an MPFF should not constitute a violation of the rule, the NRC staff determined that the licensee should have monitored the extraction steam system in accordance with paragraph (a)(1) with goals established and corrective actions taken to monitor and predict pipe wall thickness degradation. The E/C program contained deficiencies that were the cause for not detecting a degraded condition leading to a major pipe rupture. As a result, the licensee did not establish condition-monitoring performance measures under the maintenance rule that would demonstrate that the extraction steam system was maintained by an adequate preventive maintenance program. Therefore, the licensee's technical basis for tracking the extraction steam system in accordance with (a)(2) was flawed and the system should have been monitored under (a)(1) with goals and corrective actions taken to improve system performance.

- The NRC agrees with the industry that the rule is resource intensive for a variety of reasons including the following:

    (1)   Because the MR is the first performance-based, risk-informed rule, both the industry and the NRC had little experience with this new approach to regulation.

    (2)   Implementing guidance is very detailed.

    (3)   There were many different lessons learned approaches to implementing the MR that are not listed in rule implementation guidance documents.

    (4)   Licensees were given flexibility with rule implementation, which led the industry to develop diverse technical approaches to implementation.

    (5)   The industry can use lessons learned from rule implementation to improve and implement other risk-informed, performance-based regulatory programs (i.e., graded quality assurance, TS in-service testing and in-service inspection, etc.).

- The NRC staff will continue to monitor the efficacy of the rule using a variety of approaches that are still under evaluation. The staff agrees that the rule appears to be improving overall plant performance; however, more time is needed to develop appropriate methods for measuring the MR's efficacy.

Additional details on the issues discussed above were presented in an NRC memorandum from Peter A. Balmain, HQMB, to Stewart L. Magruder, PGEB, dated June 3, 1998 (NRC, 1998b). This memorandum can also be found by selecting *Public Meetings* on the NRC Maintenance Rule Home Page on the World Wide Web. See Section 2.12 for additional details.

On June 4, 1998, the NRC staff held a public meeting with the NEI, Winston & Strawn, industry representatives, and licensee personnel related to MR issues. On the basis of lessons learned from conducting the MRBIs, the NRC presented the industry with the following discussion topics that needed further clarification for proper rule implementation:

- Scope of SSCs under 50.65(b)

    − Non-safety-related SSCs not in scope that could cause safety-related SSC failures:

        ○   cathodic protection
        ○   chemical addition/chlorination

    − Support systems for in-scope EOP items
    − Emergency response computers

- Preventive maintenance analysis/review of (a)(2) SSCs
- Unavailability determination
- Taking credit for operator action

- Strengths observed during MRBIs
- Function failures
- Minimum wall thickness/leakage
- Frequency of goal monitoring
- Transition to performance-based inspection

Additional details on the issues discussed above were presented in an NRC memorandum from Peter A. Balmain, HQMB to Stewart L. Magruder, PGEB, dated June 10, 1998 (NRC, 1998c). This memorandum can also be found by selecting *Public Meetings* on the NRC Maintenance Rule Home Page on the World Wide Web. See Section 2.12 for additional details.

On July 28, 1998, the NRC staff held a public meeting with representatives from the NEI to discuss issues related to proposed 10 CFR 50.65(a)(4) rulemaking as well as issues associated with the second draft revision to NUMARC 93-01, Section 11.0, "Evaluation of Systems To Be Removed From Service," dated July 1, 1998 (NEI, 1998). Specifically, the NEI developed its draft guidance and submitted it to the NRC to address the proposed revisions to 10 CFR 50.65(a)(4) as described in the SRM dated December 17, 1997 (NRC, 1997b). This SRM directed the staff to develop proposed revisions to the MR to require licensees to perform safety assessments and take the results into account before removing SSCs from service for maintenance, subject to certain provisions delineated in the SRM. Additional details of the proposed 10 CFR 50.65(a)(4) rulemaking are found in an NRC memorandum from Richard Correia to Stewart Magruder, dated August 5, 1998 (NRC, 1998d).

The NEI recommended that the NRC clarify the rule language proposed in the SRM, which states "ensure the plant is not placed in a risk-significant configuration." Specifically, the NEI stated that the language implies a "risk value" will be quantitatively determined and that the decision metric will be used to define "risk significant." Although this approach can be used for certain assessments, it is not generally practical in state-of-the-art PRAs. The NEI also maintained that shutdown risk is not well defined, and the rule is being concurrently clarified to explicitly denote shutdown applicability. Further, the implementation history of the maintenance rule recognizes that the assessments can be performed qualitatively.

The NEI also stated that another important consideration in developing the rule revision is how it will integrate with TS requirements. The rule would appear to establish two regulatory requirements with regard to outage times for equipment covered by TS limiting conditions for operation. If the assessment requirement maintains reasonable flexibility (needed to address generally non-quantifying issues such as transition risk or operator burden), it would generally reflect current practices and the overlap with TSs is minimized. If the new requirement to address the assessment results is interpreted as a regulatory requirement based on a defined threshold (either in the rule or through inspection and enforcement), the issue of overlap with TSs is significant.

During a meeting on July 28, 1998, the NRC staff conveyed to the NEI several comments and recommendations to clarify proposed changes to the NEI guidance. These comments included the need for the safety assessments to

(1)   be completed for planned and emergent work changes to the maintenance schedule to assess real-time changes in plant configurations,

(2)   address both online and shutdown maintenance activities,

(3)   address all SSCs within the scope of 10 CFR 50.65(b) that are taken out of service for maintenance,

(4)   address both quantitative and qualitative assessments, and

(5)   not be limited to the existing Level 1 PRA models (i.e., consider Level II PRA analyses, external events, changes and updates to PRA models, etc.).

The NRC staff also determined that the quality and limitations of safety assessment tools should be addressed to assure appropriate analyses of plant configurations. The responsibilities of licensees' plant management regarding safety assessments that approve removing multiple SSCs in higher risk configurations should be addressed. The assessments must be risk informed if licensees amend their operating licensee using the process described in RG 1.177, "An Approach for Plant-Specific, Risk-Informed Decision Making: Technical Specifications." In addition, the NRC staff presented several other technical clarifications regarding the definition of unavailability; shutdown considerations before conducting maintenance; additional technical guidance needed to remove one, two, or more SSCs from service; and temporary risk guidelines.

The NRC staff discussed additional details on these issues with the NEI and reiterated this information in attachments to an NRC memorandum from Bob Latta, HQMB, to Stewart Magruder, PGEB, dated August 5, 1998 (NRC, 1998e). This memorandum can also be found by selecting *Public Meetings* on the NRC Maintenance Rule Home Page on the World Wide Web. See Section 2.12 for additional details.

On August 13, 1998, two NRC staff members attended an American Nuclear Society (ANS)1998 Utility Work Conference in Amelia, Florida, to present the NRC's perspective on MR implementation (NRC, 1998f). During the meeting, industry representatives raised questions concerning the broad INPO EPIX database definition of availability in which credit can be taken for operator actions over an unspecified time period. The industry was concerned that this definition of availability was not consistent with the NUMARC 93-01 definition of unavailability, which does not take credit for any operator action. An industry representative stated that this could create an unnecessary burden on the industry requiring licensees to maintain two databases to track unavailability (i.e., one for the maintenance rule and one for INPO EPIX). The NRC staff position was that licensees should consider the equipment available only if the licensee can return the equipment to its automatic function with *one action by a dedicated operator following a procedure*. The NRC staff reiterated to the industry that it would work with the NEI and INPO to resolve this issue in MR guidance documents.

The NRC staff also gave the industry an overview on the need for the rule; development of this first risk-informed, performance-based rule; industry flexibility in rule implementation; the results of MRBIs; lessons learned from MRBIs; variations in licensees' methods of implementing the rule; the large amount of NRC and industry resources needed to implement the rule; and proposed 10 CFR 50.65(a)(4) rulemaking to require safety assessments before taking SSCs out of service for maintenance. Additional details on the NRC staff's presentation at the ANS conference can be found on the NRC Maintenance Rule Web Page as noted in Section 2.12.

On August 27, 1998, the NRC staff held a public meeting with representatives from the NEI, Winston & Strawn, NSSS vendors, other industry service organizations and licensees regarding MR related issues. The topics of the meeting were the definition of SSC unavailability, timeliness of periodic evaluations, efficacy of the maintenance rule, and an open discussion on miscellaneous topics.

The subject of unavailability included discussions regarding the functionality of portions of the reactor protection system (RPS) as well as unavailability characterization when portions of the RPS are in the "test" position. In addition, the NRC staff discussed its position on taking credit for operator actions to declare equipment available (i.e., should be limited to one operator action using a dedicated procedure). The NRC and industry agreed that additional regulatory and industry guidance on the subject was needed to ensure a common understanding among all licensees.

Participants in the meeting also discussed periodic evaluations and the meaning of associated terms used in the rule, such as refueling cycle not to exceed 24 months, as well as concepts of balancing reliability and availability. The NRC staff noted that the quality of the evaluations was of greater importance than the timeliness.

During a broader range discussion of MR efficacy, a licensee representative stated that decreasing outage time with little or no impact to a plant's baseline CDF profile while on line would be a good indicator of MR efficacy. In response to a question from Winston & Strawn about using the number of SSCs in the (a)(1) category as an indicator, the NRC staff responded that there was nothing in the rule or the regulatory guidance that prohibits placing all SSCs in the (a)(1) category, and it was not NRC policy to use the (a)(1) monitoring category as an indicator of MR efficacy. The NRC staff also pointed out that it was not meaningful to attempt plant to plant (a)(1) comparisons; however, the attendees were reminded that a licensee was not constrained from using the SSCs in the (a)(1) category as an internal indicator of maintenance effectiveness if it so desired. The NRC staff also noted that management of the scheduling and planning process in controlling the amount of work on a plant's maintenance backlog could also be used as an indicator to monitor the effectiveness of maintenance. Given all of these suggested indicators, the NRC staff continues to evaluate MR efficacy issues.

Additional details on these issues were discussed with NEI and presented in an NRC memorandum from Ed Ford, HQMB, to Stewart Magruder, PGEB, dated September 11, 1998 (NRC, 1998g). This memorandum can also be found by selecting *Public Meetings* on the NRC Maintenance Rule Home Page on the World Wide Web. See Section 2.12 for additional details.

On February 4, 1999, the ACRS held a public meeting with the NRC staff to discuss rulemaking activities on 10 CFR 50.65(a)(4). On the basis of industry comments, the staff proposed a new revision to the proposed rulemaking on 10 CFR 50.65(a)(4) that states

> Before performing maintenance activities (including but not limited to surveillance, post-maintenance testing, corrective and preventive maintenance) on structures, systems, or components within the scope of this section, the licensee shall assess and manage any increases in risk that may result from the proposed maintenance activities.

On April 8, 1999, the ACRS held a second public meeting with NRC staff to discuss draft regulatory guide (DG) 1082, "Assessing and Managing Risk Before Maintenance Activities at Nuclear Power Plants," dated April, 1998. This draft guidance was also placed on the Maintenance Rule Home Page on the World Wide Web and in the public document room. On April 14, 1999, the ACRS sent a letter to the Chairman of the NRC (NRC, 1999a) on the proposed final revision to 10 CFR 50.65(a)(4) and DG-1082. In this letter, the ACRS determined that both the staff and industry agreed that 10 CFR 50.65 needed to require the industry to complete assessments before performing maintenance. NEI and the industry suggested that the scope of the revised rule be limited to safety significant SSCs that are modeled in PRAs and/or other risk evaluation processes and risk ranked in accordance with the guidance in NUMARC 93-01. The ACRS determined that it was not apparent that components ranked as having low safety significance will continue to be of low safety significance under all the configurations that can occur when multiple components are taken out of service. The ACRS also determined that the proposed rule and DG-1082 are sufficiently flexible that the assessments can be performed without imposing significant burden on licensees. In addition, the ACRS determined that the assessments should be required for an expanded scope of maintenance activities and during all modes of plant operation, including shutdown. On the basis of these comments, the NRC staff proposed the following rulemaking text for 10 CFR 50.65(a)(4):

> Before performing maintenance activities (including but not limited to surveillance, post-maintenance testing, corrective and preventive maintenance) on structures, systems, or components within the scope of this section, the licensee shall assess and manage any increases in risk that may result from the proposed maintenance activities. The scope of the assessment may be limited to structures, systems, and components that a risk-informed evaluation process has shown to be significant to public health and safety

On May 5, 1999, the Commission held a public meeting with the NRC staff, the NEI, and industry stakeholders to discuss the final rulemaking language in 10 CFR 50.65(a)(4) and implementing guidance in DG-1082. On the basis of comments from NRC, NEI, and the industry, the Commission approved the final rulemaking language in 10 CFR 50.65(a)(4). The Commission also directed the staff to complete the final rulemaking package by May 17, 1999, for Commission review and approval. Details on comments from the public and the industry on the proposed rule, the final rule, and the regulatory analysis are presented in SECY-99-133, "Final Revision to 10 CFR 50.65 to Require Licensees to Perform Assessments Before Performing Maintenance," dated May 17, 1999.

On May 6, 1999, the NRC staff met with the ACRS to discuss final rulemaking in 10 CFR 50.65(a)(4) and the DG-1082 implementing guidance. The ACRS approved final rulemaking for 10 CFR 50.65(a)(4) with comments on the implementing guidance. The ACRS determined that the NRC staff should issue the revised RG 1.160, Revision 3, for industry use before implementing the revised rule. In addition, the NRC staff should hold a workshop with the industry on the revised guidance. The rule and RG 1.160 should become effective 120 days after RG 1.160 is published.

On June 18, 1999, the Commission issued an SRM approving the final rule language noted in the staff recommendations in SECY-99-133. The Commission directed the NRC staff to amend 10 CFR 50.65 to require that, before performing maintenance activities, power reactor licensees shall assess and manage the increase in risk that may result from performing maintenance activities. The Commission also directed the staff to complete the regulatory guidance for NRC licensees and inspectors that will not become effective until the regulatory guide is in place for 120 days. Before issuing the final regulatory guide, the NRC staff should send it to the Commission for review and approval.

Conclusion

The NRC staff concludes that conducting public meetings and workshops with industry stakeholders was an effective method of reaching a common understanding regarding technical issues related to proper implementation of the maintenance rule.

Recommendations

The NRC staff should continue to hold public meetings and workshops with industry stakeholders to revise MR related guidance documents to provide for consistent MR implementation guidance. This includes revising RG 1.160 to address 10 CFR 50.65(a)(4) rulemaking activities and other technical issues associated with proper implementation of the rule.

2.12 NRC Maintenance Rule Home Page on the World Wide Web

As part of the NRC staff's continuing efforts to communicate with the stakeholders, the NRC issued Administrative Letter 98-01, dated February 20, 1998 (NRC, 1998a), to announce the availability of the maintenance rule Web site on NRC's External Home Page at:

http://www.nrc.gov/NRR/mrule/mrhome.htm

The maintenance rule Web site is also accessible from the Nuclear Reactors Home Page at:

http://www.nrc.gov/reactors.html

When accessing the Web site, the user must ensure that the uniform resource locator (URL) is typed exactly as shown because the Web server file name is case sensitive. Anyone with questions concerning information on this Web site should call Donnie Ashley, NRR, at (301) 415-3191 or e-mail dja1@nrc.gov, or mrule@nrc.gov.

This interactive site contains MR-related documents, including 10 CFR 50.65, regulatory and industry guides, inspection procedures, selected SECY papers, the SOCs, and summaries of various meetings with external groups. The MR pilot and baseline inspection reports are also included on the site with links to the Office of Enforcement Home Page for those licensees that have been the subject of escalated enforcement. A fill-in-the-blank form is available to ask questions of the staff concerning specific MR implementation issues and receive direct answers. As of September 1999, the staff has answered more than 120 questions submitted via the Web site. A "What's New" page lists a chronology of the Web site and describes recent changes. A recent innovation is the addition of the "mailing list," which contains the e-mail addresses of individuals who have requested that they be notified when the Web site changes or new information is added. The user has the ability to download a compressed copy of any baseline inspection report on the sites inspected.

Conclusion

The MR Home Page is an effective method of communicating NRC staff positions on MR implementation issues.

Recommendation

The NRC staff should continue to answer industry questions on MR implementation issues and update the MR Home Page when new information becomes available.

3 OVERALL CONCLUSIONS

The MRBI teams concluded that the requirements of 10 CFR 50.65 can be met using NUMARC 93-01 as endorsed by RG 1.160; however, some weaknesses in these guidance documents were noted. While conducting the MRBIs, the NRC revised RG 1.160 to address most of these weaknesses. The rulemaking on 10 CFR 50.65(a)(4) approved by the Commission on June 18, 1999, along with revisions to RG 1.160 and NUMARC 93-01, should also improve MR efficacy and result in continued overall improvement in industry performance in the area of maintenance. Inspection Procedure 62706 was determined to be adequate to evaluate licensees' MR programs, but will also need to be revised to reflect the rulemaking on 10 CFR 50.65(a)(4).

Generally, licensees did a good job of identifying SSCs within the scope of the MR. On average, the MRBIs identified two to five additional SSCs or SSC functions per site that should have been included within the scope of the MR.

Since the maintenance rule is a performance-based regulation, licensees have flexibility to add or remove SSCs if an adequate technical basis existed for including or excluding the SSC in question. In addition, licensees can exclude a non-safety-related SSC if its performance demonstrates that it did not contribute to events that require non-safety-related SSCs to be included within the scope of the rule as defined under 10 CFR 50.65(b)(2).

The methods used by licensees to establish risk significance were consistent with the guidance in NUMARC 93-01. The use of an expert panel review process to integrate deterministic considerations with PRA or IPE risk insights is an appropriate and practical method of determining SSC risk significance. At most sites, the expert panel members were knowledgeable of the MR and had extensive industry experience to support the decisionmaking process for determining safety (risk) significant SSCs. The required composition of the expert panels was consistent with the guidance presented in NUMARC 93-01.

In some instances, licensees did not (1) establish adequate goals commensurate with safety, (2) provide adequate technical bases for performance criteria (measures) which demonstrate effective preventive maintenance, and (3) statistically link goals or performance measures to the assumptions in the PRA. Licensees also did not adequately monitor or track FFs, MPFFs, repetitive MPFFs, and/or unavailability for some HSS and LSS standby SSCs. Some licensees' root cause or cause determination analyses and corrective actions were inadequate. The NRC concluded that licensees' inadequate monitoring was, in part, due to a reluctance to move SSCs into the paragraph (a)(1) monitoring category or a lack of awareness of goals or performance measures being insufficient or exceeded.

The perceived reluctance of licensees to place SSCs in the paragraph (a)(1) monitoring program was effectively addressed through guidance provided in RG 1.160, Revision 2, dated March, 1997, and Information Notice (IN) 97-18, "Problems Identified During Maintenance Rule Baseline Inspections," dated April 14, 1997.

OVERALL CONCLUSIONS

The NRC found that the industry generally did an adequate job of developing appropriate monitoring and tracking programs for SSCs monitored under paragraphs (a)(1) and (a)(2). Most licensees took safety into consideration by determining risk and evaluating whether goals or performance measures should be at the system, train, or component level. However, some licensees did not adequately consider the reliability and availability assumptions in their PRAs. The NRC also found that some licensees should monitor HSS I&C systems at the train and/or channel level (i.e., RPS, ESFAS). During the initial MRBIs, some licensees did not establish adequate MR monitoring programs for structures; however, this was effectively resolved when the NRC issued guidance in RG 1.160, Revision 2. Although not a requirement of the MR, licensees are encouraged to monitor and/or track FFs as well as MPFFs.

The MRBI teams found that most licensees established appropriate methods for balancing reliability and availability. A few licensee used conditional probability as a PRA method to establish bounding limits between reliability and availability. These licensees monitored and/or tracked HSS SSC reliability and availability data separately and used the conditional probability value to statistically determine whether a balance was being achieved between reliability and availability. The NRC found this to be acceptable; however, licensees should not use conditional probability to mask reliability and availability on HSS SSCs. In addition, a few licensees did not complete their periodic evaluations on time. Many licensees also used the periodic evaluations required by 10 CFR 50.65(a)(3) to determine the overall effectiveness of their MR implementation programs and made adjustments where necessary to improve plant performance.

The NRC staff concluded that 10 CFR 50.65 should be revised to require licensees to perform safety assessments before taking equipment out of service for maintenance. The NRC staff proposed rulemaking on a new 10 CFR 50.65(a)(4) section in accordance with the SRM to SECY 97-173 (NRC, 1997b). Specifically, the NRC staff proposed a slight revision to the SRM language in which the last sentence in 10 CFR 50.65(a)(4) would be revised to state, "The results of this assessment shall be used to ensure that the plant is not placed in risk-significant configurations or configurations that would degrade the performance of safety functions to an unacceptable level." The industry commented that new terms (i.e., risk significant configurations) could not be easily defined given the current state of the art of and differences in PRAs. The NRC staff proposed a new revision to 10 CFR 50.65(a)(4) to address this concern. The new revision to 10 CFR 50.65(a)(4) is quoted in Sections 2.7.2 and 2.11.1 of this report. The Commission approved the final rule with the new paragraph (a)(4) on June 18, 1999 (NRC, 1999c). The NRC and industry will revise their guidance documents, accordingly, and NRC will revise its inspection procedures.

In general, licensees' self-assessments were effective at identifying weaknesses and areas needing improvement in MR programs. The MRBI results typically characterized the self assessments as a strength in MR program implementation.

The NRC completed only three MRBIs at decommissioning status plants: Zion, Big Rock Point, and LaCrosse BWR. With the exception of one plant (LaCrosse BWR), the NRC concluded that licensees developed successful methods for implementing the MR for plants in a decommissioning status. Although the population of SSCs within the scope of the MR is much smaller for plants in a decommissioning status, many of the methods used to monitor the performance or condition of SSCs needed to maintain the spent fuel are similar to the methods described in NUMARC 93-01 (i.e., monitor reliability, availability, or condition) for an operating plant.

The NRC used an MR enforcement panel to provide for consistent inspection and enforcement of the rule while conducting MRBIs. The enforcement panel consisted of a regional branch chief, the HQMB branch chief and section chief responsible for oversight of the MR, the MRBI team leader, and an enforcement specialist from the Office of Enforcement (OE). The NRC concluded that the enforcement panel was an effective method of maintaining inspection and enforcement consistency for all MRBIs. In addition, the NRR staff support members who participated in all of the MRBIs helped MRBI teams maintain inspection and enforcement consistency.

The NRC staff concluded that public meetings conducted with industry stakeholders provided effective methods of establishing a common understanding on technical issues related to proper implementation of the rule. In addition, the MR Home Page on the World Wide Web was determined to be an effective method of communicating staff positions regarding MR implementation issues submitted by the industry.

The MRBI teams concluded that this risk-informed, performance-based approach to implementing the rule is practical but resource intensive for both the NRC and the industry. The lessons learned from this approach should be used in all other risk-informed, performance-based approaches to reduce regulatory burden in other areas.

4 RECOMMENDATIONS

Since the maintenance rule is a risk-informed, performance-based regulation, licensees have the flexibility to add or remove SSCs from the scope of the rule if an adequate technical basis exists for including or excluding the SSC in question. Licensees can exclude an SSC if its performance demonstrates that it did not contribute to events that cause non-safety-related SSCs to be included within the scope of the rule as defined under 10 CFR 50.65(b)(2). Using the flexibility of rule implementation guidance, licensees can determine adequate technical bases for making performance-based decisions on the scope of SSCs within 10 CFR 50.65(b).

As a result of the MRBIs, the NRC staff identified the following issues and recommendations concerning the risk significant determination process. The following recommendations should be considered when clarifying the related guidance in NUMARC 93-01:

- NUMARC 93-01, Sections 9.3.1.1 and 9.3.1.2, recommended that licensees eliminate RRW measures and cut sets that are not related to maintenance (e.g., operator error, external and initiating events) from the risk-determination process. This guidance was provided to avoid the potential masking of the importance of certain SSCs due to PRA modeling assumptions (e.g., higher probability estimates) used for operator errors, and external and internal events. In a few cases, this guidance caused licensees to identify some SSCs as LSS rather than HSS. RRW measures and cut sets should be removed cautiously to ensure that the eliminated cut sets do not implicitly account for maintenance activities associated with certain SSCs that have high importance ranking.

- Plant-specific PRA models used for risk-importance analyses should be of sufficient quality to ensure consistent results in the SSC safety-significance categorizations. In determining safety significance of SSCs, all licensees performed importance analyses using PRA models that were developed for IPE and/or IPEEE documents. The quality of the PRAs used at most of the licensee sites has not been assessed through peer or industry reviews. For MRBI purposes, the PRA was assumed to have sufficient quality to support the risk categorization process since there was an expert panel process in place to compensate for limitations in the PRA. This issue on PRA quality is addressed in ongoing NRC and industry initiatives to identify requirements for PRA standards in a PRA certification process.

- For future risk-informed performance-based inspection initiatives, licensees' risk determination methodologies for establishing HSS and LSS SSCs within the scope of other regulatory applications should also take into account the plant specific baseline risk level. This could result in more suitable importance measure thresholds being used for the specific plant and regulatory applications under review.

- NRC and licensee resource commitments for risk-informed, performance-based regulatory activities are initially high, and should be acknowledged and committed to up front. Risk-informed, performance-based regulatory activities require coordination with the industry to develop explicit implementation guidance.

- At several sites, the PRA modeling assumptions and data were not updated to reflect the "as-operated" plant configuration and operational experience. The importance measure calculations based on these PRA models would produce inconsistent risk ranking results. Thus, the risk ranking results should be re-evaluated whenever (1) major plant design changes are implemented, (2) the PRA models are updated, and (3) new reliability and availability data become available.

- Currently, NUMARC 93-01, Section 9.3.1, provides guidance only on the use of an expert panel process to determine the safety significance of SSCs by evaluating PRA risk ranking results in conjunction with the design and operating experience considerations. The industry may also consider using the expert panel to support decision making activities associated with other parts of the rule such as the scope of the MR, establishment of goals and performance measures, when SSCs should be moved from paragraph (a)(2) to (a)(1) or from paragraph (a)(1) back to (a)(2), corrective actions to improve SSC performance, and the periodic evaluations under paragraph (a)(3) of the rule.

Even if individual system or system train level performance is acceptable, licensees should monitor similar components (i.e., circuit breakers, motorized operated valves, solenoid operated valves, limit switches, relays, etc.) across system or train boundaries under paragraph (a)(1) when common-mode repetitive MPFFs occur.

The industry may wish to consider additional methods (e.g., MPRFFs) to monitor the effectiveness of maintenance on redundant equipment (i.e., instrument and service air compressors, service water pumps, etc.). This could eliminate the need to monitor unavailability for trains of equipment that are removed from service for extended periods of time. All licensees should develop performance measures that monitor HSS instrumentation and control systems at the train or channel level or both. In addition to SSC MPFFs that cause automatic scrams, the industry should also count SSC MPFFs that cause manual scrams. Licensees are encouraged to monitor both FFs and MPFFs, but this is not a requirement of the rule.

Licensees may use conditional probability methods to determine the appropriate bounding limits for reliability and availability, as long as performance measures for reliability and availability are established separately. If used properly, this method can verify if a balance is being achieved between reliability and availability to meet the requirements of 10 CFR 50.65(a)(3). Licensees can continue to use the paragraph (a)(3) periodic evaluations to determine the overall efficacy of their MR implementation programs and to make adjustments where necessary to improve plant performance.

Licensees should continue self-assessment activities periodically to verify that MR programs do not lapse and remain as living programs that demonstrate the effectiveness of maintenance activities in improving overall plant performance.

Licensees should continue to use the self-assessments and paragraph (a)(3) periodic evaluations to review the performance of SSCs to ensure that appropriate goal setting, monitoring, and preventive maintenance are maintained and to monitor the overall efficacy of their MR program. Licensees may choose the number of SSCs in the paragraph (a)(1) monitoring category as an internal indicator on the efficacy of the MR at their plants; however, the NRC does not consider the number of SSCs in the paragraph (a)(1) monitoring category to be an indicator of ineffective maintenance. The NRC may consider the number of SSCs with repetitive returns to the paragraph (a)(1) monitoring category as another indicator of MR efficacy.

The NRC should work with industry stakeholders to complete industry guidance that gives standard methods for implementing the rule at plants in a decommissioning status. This guidance should be endorsed by RG 1.160.

The NRC should continue to use the MR enforcement panel to maintain enforcement consistency for all MR issues identified by regional inspectors. The NRC staff should also revise the EGM on MR enforcement to incorporate lessons learned from the MRBIs.

The NRC staff should continue to hold public meetings and workshops with industry stakeholders to revise MR related documents to present consistent MR implementation guidance. This includes revising RG 1.160, and IP 62706 and 62707 to address 10 CFR 50.65(a)(4) and other technical issues associated with proper implementation of the rule. The NRC staff shall also update inspection procedures to reflect changes in the rule. In addition, the NRC staff shall update the NRC Maintenance Rule Home Page as new information becomes available.

5 REFERENCES

DG-1051   U.S. Nuclear Regulatory Commission, "Monitoring the Effectiveness of Maintenance at Nuclear Power Plants," RG 1.160, Revision 2, August 1996

DG-1082   U.S. Nuclear Regulatory Commission, "Assessing and Managing Risk Before Maintenance Activities at Nuclear Power Plants," Draft Regulatory Guide 1082, April 1999

EPRI, 1995   Electric Power Research Institute, "PSA Applications Guide," Topical Report 105396, 1995

IN 97-18   U.S. Nuclear Regulatory Commission, " Problems Identified During Maintenance Rule Baseline Inspections," Information Notice 97-18, April 14, 1997

IP 62002   U.S. Nuclear Regulatory Commission, "Inspection of Structures, Passive Components, and Civil Engineering Features at Nuclear Power Plants," Inspection Procedure 62002, December 31, 1996

IP 62706   U.S. Nuclear Regulatory Commission, "The Maintenance Rule," NRC Inspection Procedure 62706, Revision 1, December 31, 1997

IP 62707   U.S. Nuclear Regulatory Commission, "Maintenance Observations," Inspection Procedure 62707, April 22, 1996

NEI, 1996   Nuclear Energy Institute (NEI) 96-03, "Guideline for Monitoring the Condition of Structures at Nuclear Power Plants," December 1996

NEI, 1997a   Nuclear Energy Institute (NEI), "10 CFR 50.65 Maintenance Rule Implementation for Shutdown Plants," Industry White Paper, August 1997

NEI, 1997b   Letter from Nuclear Energy Institute to NRC Executive Director for Operations, October 10, 1997

NEI, 1998   Nuclear Energy Institute, "Evaluation Of Systems To Be Removed From Service," Draft revision to NUMARC 93-01, Section 11, July 1, 1998

NRC, 1991   U.S. Nuclear Regulatory Commission, "Statements of Consideration for 10 CFR 50.65" (56 FR 31306), July 10, 1991

NRC, 1997a   U.S. Nuclear Regulatory Commission, Memorandum from Suzanne C. Black to David Matthews, January 17, 1997

NRC, 1997b   U.S. Nuclear Regulatory Commission, "Staff Requirements Memorandum in Response to SECY 97-173," December 17, 1997

NRC, 1998a   U.S. Nuclear Regulatory Commission, Administrative Letter 98-01, February 20, 1998

NRC, 1998b   U.S. Nuclear Regulatory Commission, Memorandum from Peter Balmain to Stewart Magruder, June 3, 1998

NRC, 1998c   U.S. Nuclear Regulatory Commission, Memorandum from Peter Balmain to Stewart Magruder, June 10, 1998

NRC, 1998d   U.S. Nuclear Regulatory Commission, Memorandum from Richard Correia to Stewart Magruder, August 5, 1998

NRC, 1998e   U.S. Nuclear Regulatory Commission, Memorandum from Robert Latta to Stewart Magruder, August 5, 1998

NRC, 1998f   U.S. Nuclear Regulatory Commission, "NRC Staff Presentation on Maintenance Rule Implementation Issues," by Richard Correia and John D. Wilcox, to the ANS Utility Working Group Conference, Amelia, Florida, August 13, 1998 (Go to Maintenance Rule Home Page.)

NRC, 1998g   U.S. Nuclear Regulatory Commission, Memorandum from Edward Ford to Stewart Magruder, September 11, 1998

NRC, 1999a   U.S. Nuclear Regulatory Commission, Advisory Committee on Reactor Safeguards (ACRS), Memorandum from Dana A. Powers, Chairman, ACRS, to Shirley A. Jackson, Chairman, NRC, April 14, 1999

NRC, 1999b   U.S. Nuclear Regulatory Commission, "Final Revision to 10 CFR 50.65 to Require Licensees to Perform Safety Assessments Before Performing Maintenance," Commission Voting Record, June 18, 1999

NRC, 1999c   U.S. Nuclear Regulatory Commission, Staff Requirements Memorandum in Response to SECY 99-133, June 18, 1999

NUMARC 91-06   Nuclear Management and Resources Council, Inc., "Guidelines for Industry Actions to Assess Shutdown Management," NUMARC 91-06, Vol. 2, December 1991

NUMARC 93-01   Nuclear Management and Resources Council, Inc., "Industry Guideline for Monitoring the Effectiveness of Maintenance at Nuclear Power Plants," NUMARC 93-01, Revision 2, April 1996

NUREG-1526   U.S. Nuclear Regulatory Commission, "Lessons Learned From Early Implementation of the Maintenance Rule at Nine Nuclear Power Plants," NUREG-1526, C. Petrone, R. Correia, and S. Black, June 1995

REFERENCES

NUREG-1605	U.S. Nuclear Regulatory Commission,"Methodology of Plant Configurations and Pilot Applications: Lessons Learned," NUREG-1605, January 1999
NUREG/CR-5424	U.S. Nuclear Regulatory Commission, "Eliciting and Analyzing Expert Judgment," NUREG/CR-5424, January 1990
NUREG/CR-5500(2)	U.S. Nuclear Regulatory Commission, "Reliability Study: Westinghouse Reactor Protection System, 1984–1995," NUREG/CR-5500, Volume 2, April 1999
NUREG/CR-5500(3)	U.S. Nuclear Regulatory Commission, "Reliability Study, General Electric Reactor Protection System, 1984–1995," NUREG/CR-5500, Volume 3, February 1999
RG 1.160(1)	U.S. Nuclear Regulatroy Commission, "Monitoring the Effectiveness of Maintenance at Nuclear Power Plants," Regulatory Guide 1.160, Revision 1, January 1995
RG 1.160(2)	U.S. Nuclear Regulatory Commission,"Monitoring the Effectiveness of Maintenance at Nuclear Power Plants," Regulatory Guide 1.160, Revision 2, March 1997
RG 1.174	U.S. Nuclear Regulatory Commission, "An Approach for Using Probabilistic Risk Assessment in Risk-Informed Decisions on Plant-Specific Changes to the Licensing Basis," Regulatory Guide 1.174, July 1998
RG 1.177	U.S. Nuclear Regulatory Commission, "An Approach for Plant-Specific, Risk-Informed Decision Making: Technical Specifications," Regulatory Guide 1.177, September 1998
SECY-97-055	U.S. Nuclear Regulatory Commission, "Maintenance Rule Status, Results, and Lessons Learned," SECY-97-055, March 4, 1997
SECY-97-173	U.S. Nuclear Regulatory Commission, "Potential Revision to 10 CFR 50.65(a)(3) of the Maintenance Rule to Require Licensees to Perform Safety Assessments," SECY-97-173, August 1, 1997
SECY-98-165	U.S. Nuclear Regulatory Commission, "Proposed Revision to 10 CFR 50.65(a)(3) to Require Licensees to Perform Safety Assessments," SECY-98-165, July 2, 1998
SECY-99-133	U.S. Nuclear Regulatory Commission, "Final Revision to 10 CFR 50.65 to Require Licensees to Perform Safety Assessments Before Performing Maintenance," SECY 99-133, May 17, 1999

APPENDIX A

THE MAINTENANCE RULE

The Maintenance Rule, effective on July 10, 1996

§50.65 Requirements for monitoring the effectiveness of maintenance at nuclear power plants.

(a)(1) Each holder of a license to operate a nuclear power plant under §§ 50.21(b) and 50.22 shall monitor the performance or condition of structures, systems, or components, against licensee-established goals, in a manner sufficient to provide reasonable assurance that such structures, systems, and components, as defined in paragraph (b), are capable of fulfilling their intended functions. Such goals shall be established commensurate with safety and, where practical, take into account industry-wide operating experience. When the performance or condition of a structure, system, or component does not meet established goals, appropriate corrective action shall be taken.

(2) Monitoring as specified in paragraph (a)(1) of this section is not required where it has been demonstrated that the performance or condition of a structure, system, or component is being effectively controlled through the performance of appropriate preventive maintenance, such that the structure, system, or component remains capable of performing its intended function.

(3) Performance and condition monitoring activities and associated goals and preventative maintenance activities shall be evaluated at least every refueling cycle provided the interval between evaluations does not exceed 24 months. The evaluations shall be conducted taking into account, where practical, industry-wide operating experience. Adjustments shall be made where necessary to ensure that the objectives of preventing failures of structures, systems, and components through maintenance is appropriately balanced against the objective of minimizing unavailability of structures, systems, and components due to monitoring or preventative maintenance. In performing monitoring and preventative maintenance activities, an assessment of the total plant equipment that is out of service should be taken into account to determine the overall affect on performance of safety functions.

(b) The scope of the monitoring program specified in paragraph (a)(1) of this section shall include safety related and nonsafety-related structures, systems, and components as follows:

(1) Safety-related structures, systems, and components that are relied upon to remain functional during and following design basis events to ensure the integrity of the reactor coolant pressure boundary, the capability to shutdown the reactor and maintain it in a safe shutdown condition, and the capability to prevent or mitigate the consequences of accidents that could result in potential offsite exposure comparable to the guidelines in §50.34(a)(1) or §100.11 of this chapter, as applicable.

(2) Nonsafety-related structures, systems, or components:

(i) That are relied upon to mitigate accidents or transients or are used in emergency operating procedures (EOPs); or

(ii) Whose failure could prevent safety-related structures, systems, and components from fulfilling their safety-related function; or

(iii) Whose failure could cause a reactor scram or actuation of a safety-related system.

(c) The requirements of this section shall be implemented by each licensee no later than July 10, 1996.

The Maintenance Rule as Amended on August 28, 1996
(This revision added requirements for plants in a decommissioning status.)

§50.65 Requirements for monitoring the effectiveness of maintenance at nuclear power plants.

(a)(1) Each holder of a license to operate a nuclear power plant under §§ 50.21(b) and 50.22 shall monitor the performance or condition of structures, systems, or components, against licensee-established goals, in a manner sufficient to provide reasonable assurance that such structures, systems, and components, as defined in paragraph (b), are capable of fulfilling their intended functions. Such goals shall be established commensurate with safety and, where practical, take into account industry-wide operating experience. When the performance or condition of a structure, system, or component does not meet established goals, appropriate corrective action shall be taken. *For a nuclear power plant for which the licensee has submitted the certifications specified in 50.82(a)(1), this section only shall apply to the extent that the licensee shall monitor the performance or condition of all structures, systems, and components associated with the storage, control, and maintenance of spent fuel in a safe condition, in a manner sufficient to provide reasonable assurance that such structures, systems, and components are capable of fulfilling their intended safety functions.*

(2) Monitoring as specified in paragraph (a)(1) of this section is not required where it has been demonstrated that the performance or condition of a structure, system, or component is being effectively controlled through the performance of appropriate preventive maintenance, such that the structure, system, or component remains capable of performing its intended function.

(3) Performance and condition monitoring activities and associated goals and *preventive* maintenance activities shall be evaluated at least every refueling cycle provided the interval between evaluations does not exceed 24 months. The evaluations shall be conducted taking into account, where practical, industry-wide operating experience. Adjustments shall be made where necessary to ensure that the objectives of preventing failures of structures, systems, and components through maintenance is appropriately balanced against the objective of minimizing unavailability of structures, systems, and components due to monitoring or preventative maintenance. In performing monitoring and *preventive* maintenance activities, an assessment of the total plant equipment that is out of service should be taken into account to determine the overall affect on performance of safety functions.

(b) The scope of the monitoring program specified in paragraph (a)(1) of this section shall include safety related and nonsafety-related structures, systems, and components as follows:

(1) Safety-related structures, systems, and components that are relied upon to remain functional during and following design basis events to ensure the integrity of the reactor coolant pressure boundary, the capability to shutdown the reactor and maintain it in a safe shutdown condition, and the capability to prevent or mitigate the consequences of accidents that could result in potential offsite exposure comparable to the guidelines in §50.34(a)(1) or §100.11 of this chapter, as applicable.

(2) Nonsafety-related structures, systems, or components:

(i) That are relied upon to mitigate accidents or transients or are used in emergency operating procedures (EOPs); or

(ii) Whose failure could prevent safety-related structures, systems, and components from fulfilling their safety-related function; or

(iii) Whose failure could cause a reactor scram or actuation of a safety-related system.

(c) The requirements of this section shall be implemented by each licensee no later than July 10, 1996.

The Maintenance Rule as Amended on June 18, 1999
(This revision added requirements for licensees to perform safety assessments
before conducting maintenance.)

§50.65 Requirements for monitoring the effectiveness of maintenance at nuclear power plants.

The requirements of this section are applicable during all conditions of plant operation, including normal shutdown operations.[1]

(a)(1) Each holder of a license to operate a nuclear power plant under §§ 50.21(b) and 50.22 shall monitor the performance or condition of structures, systems, or components, against licensee-established goals, in a manner sufficient to provide reasonable assurance that such structures, systems, and components, as defined in paragraph (b), are capable of fulfilling their intended functions. Such goals shall be established commensurate with safety and, where practical, take into account industry-wide operating experience. When the performance or condition of a structure, system, or component does not meet established goals, appropriate corrective action shall be taken. For a nuclear power plant for which the licensee has submitted the certifications specified in 50.82(a)(1), this section only shall apply to the extent that the licensee shall monitor the performance or condition of all structures, systems, and components associated with the storage, control, and maintenance of spent fuel in a safe condition, in a manner sufficient to provide reasonable assurance that such structures, systems, and components are capable of fulfilling their intended safety functions.

(2) Monitoring as specified in paragraph (a)(1) of this section is not required where it has been demonstrated that the performance or condition of a structure, system, or component is being effectively controlled through the performance of appropriate preventive maintenance, such that the structure, system or component remains capable of performing its intended function.

(3) Performance and condition monitoring activities and associated goals and preventive maintenance activities shall be evaluated at least every refueling cycle provided the interval between evaluations does not exceed 24 months. The evaluations shall be conducted taking into account, where practical, industry-wide operating experience. Adjustments shall be made where necessary to ensure that the objectives of preventing failures of structures, systems, and components through maintenance is appropriately balanced against the objective of minimizing unavailability of structures, systems, and components due to monitoring or preventative maintenance.

(4) Before performing maintenance activities (including but not limited to surveillance, post maintenance testing, corrective and preventive maintenance) on structures, systems, or components within the scope of this section, the licensee shall assess and manage any increases in risk that may result from the proposed maintenance activities. The scope of the assessment may be limited to structures, systems, and components that a risk-informed evaluation process has shown to be significant to public health and safety.[1]

(b) The scope of the monitoring program specified in paragraph (a)(1) of this section shall include safety related and nonsafety-related structures, systems, and components as follows:

(1) Safety-related structures, systems, and components that are relied upon to remain functional during and following design basis events to ensure the integrity of the reactor coolant pressure boundary,

---

[1]The amendment to this rule will become effective 120 days after Regulatory Guide 1.160, Revision 3, is issued.

the capability to shutdown the reactor and maintain it in a safe shutdown condition, and the capability to prevent or mitigate the consequences of accidents that could result in potential offsite exposure comparable to the guidelines in §50.34(a)(1) or §100.11 of this chapter, as applicable.

(2) Nonsafety-related structures, systems, or components:

(i) That are relied upon to mitigate accidents or transients or are used in emergency operating procedures (EOPs); or

(ii) Whose failure could prevent safety-related structures, systems, and components from fulfilling their safety-related function; or

(iii) Whose failure could cause a reactor scram or actuation of a safety-related system.

(c) The requirements of this section shall be implemented by each licensee no later than July 10, 1996

APPENDIX B

BACKGROUND

B.1   Objective

This appendix summarizes the lessons learned from the U.S. NRC's maintenance rule visits to pilot sites conducted between September 1994 and March 1995. The objectives of these MR pilot site visits were to review the adequacy of licensees' implementation of requirements to comply with the rule, and to discuss development of NUMARC 93-01, "Industry Guideline for Monitoring the Effectiveness of Maintenance at Nuclear Power Plants," as endorsed by RG1.160, "Monitoring the Effectiveness of Maintenance at Nuclear Power Plants," and Inspection Procedure 62706, "The Maintenance Rule." The rule, which was published on July 10, 1991, as 10 CFR 50.65, "Requirements for monitoring the effectiveness of maintenance at nuclear power plants," took effect on July 10, 1996. As revealed during these site visits, all licensees used the guidance in NUMARC 93-01 to develop and implement adequate programs with some exceptions, which were found to be acceptable in most cases. The results of these reviews were documented in NUREG-1526, "Lessons Learned from Early Implementation of the Maintenance Rule at Nine Nuclear Power Plants." Accordingly, licensees should consider the information in NUREG 1526 when making changes to their MR programs.

B.2   Need for the Maintenance Rule

In the statements of consideration for the rule, the Commission stated that such a rule is needed because effective maintenance is clearly linked to safety. Good maintenance helps limit the number of transients and challenges to safety systems by ensuring the reliability and availability of safety equipment. Good maintenance is also important in minimizing failures of non-safety-related SSCs that could initiate or adversely affect a transient or accident. Minimizing challenges to safety systems is consistent with the Commission's defense-in-depth philosophy. Additionally, maintenance is important to ensure that design assumptions and margins in the original design basis are maintained or at least are not unacceptably degraded. Therefore, the Commission concluded that maintenance at nuclear power plants is clearly important in protecting the public health and safety.

The results of the Commission's maintenance team inspections (MTIs) conducted between 1988 and 1990 indicated that licensees had generally adequate maintenance programs and were improving the implementation of their programs. However, the inspections revealed some common maintenance-related weaknesses, such as inadequate root cause analysis leading to repetitive failures; lack of equipment performance trending; and lack of consideration of plant risk in the prioritization, planning, and scheduling of maintenance. The Commission believes that the effectiveness of maintenance must be assessed continually to assure that key SSCs are capable of performing their intended functions. Further, licensees need to consider revising programmatic requirements where poor assessment results indicate ineffective maintenance.

Despite significant industry accomplishments in maintenance program content and implementation, plant events caused by the degradation or failure of plant equipment continued to occur as a result of instances of ineffective maintenance. Additionally, operational events were exacerbated by, or resulted from, plant equipment being unavailable because of maintenance activities. Most of the existing requirements and industry maintenance initiatives did not call for licensees to routinely assess the availability of safety significant SSCs. These events and circumstances attest to the need to continually assess the results of maintenance effectiveness by gathering data on equipment reliability and availability.

The Commission also acknowledged the need to broaden its ability to take timely enforcement action where maintenance activities fail to give reasonable assurance that safety-significant SSCs are capable of performing their intended functions. The Commission concluded that it was necessary to include requirements for corrective actions to address instances of ineffective maintenance, and for licensees to use the results of monitoring and assessment to improve their maintenance programs.

In addition to the preceding considerations, the Commission's conclusion that a rule requiring that the effectiveness of maintenance be monitored is predicated on the fact that the Commission's current regulations, regulatory guidance, and licensing practices do not clearly define the Commission's expectations with regard to ensuring the continued effectiveness of maintenance programs at nuclear power plants. There was no guidance regarding the monitoring of maintenance effectiveness.

Accordingly, the requirements and guidance for monitoring maintenance effectiveness and for taking corrective action when maintenance is ineffective were intended to enhance the Commission's capability to take timely and effective action against licensees with inadequate or poorly conducted maintenance to ensure prompt resumption of effective maintenance activities.

On July 10, 1991, the Commission published the final rule, 10 CFR 50.65, in the *Federal Register*. When the rule took effect on July 10, 1996, it required all nuclear power plant licensees to monitor the effectiveness of their maintenance activities. The rule provides for continued emphasis on the defense-in-depth principle by including selected non-safety-related SSCs, integrates risk consideration into the maintenance process, establishes an enhanced regulatory basis for inspection and enforcement of certain non-safety-related SSC maintenance-related issues, and gives a strengthened regulatory basis for ensuring that the progress achieved is sustained in the future.

B.3   Process-Oriented Versus Risk-Informed, Results-Oriented Regulations

Although they are not strictly defined, the terms "process-oriented" (or programmatic, or prescriptive) and "results-oriented" (or results-based, or performance-based) are generally used to describe various rulemaking activities. The processes used to implement the MR are intended to be results-oriented (i.e., results-based or performance-based) which indicates that results-oriented approaches are increasingly being used by the NRC to describe various rulemaking activities.

A process-oriented rule, the traditional approach to most rulemaking, included detailed requirements or instructions. The advantage to such rules was that they were easier to enforce because the requirements for implementing the rule are delineated in greater detail than would be the case in results-oriented rules. Using a process-oriented rule, licensees generally have a clearer idea of what they need to do to implement the requirements of the rule, and NRC inspectors have a clearer idea of what to inspect. The disadvantage to such rules is that they tend to be somewhat inflexible, and thus prevent licensees from using the most efficient and effective means of implementing the rule. Two examples of process-oriented rules are 10 CFR Part 50, Appendix J[1], "Primary Containment Leakage Testing," and Appendix R, "Fire Protection Program for Nuclear Power Plants." These rules contain detailed requirements for test frequency, test pressures, training, and record-keeping.

A results-oriented (i.e., results-based or performance-based) rule describes, in general terms, the expected results, while leaving the details of achieving those results up to the licensee. Such a rule has the advantage of allowing licensees to devise the most effective and efficient means of achieving the results described in the rule. It also allows licensees to consider safety (risk) significance when developing their programs. The disadvantage of a results-oriented rule is that it may be difficult to enforce because the requirements for compliance are less clearly defined than the requirements of a process-oriented rule. The maintenance rule, 10 CFR 50.65, is a risk-informed, results-oriented[2] rule.

Although licensees clearly prefer results-oriented regulations over process-oriented regulations, the lack of detail in such regulations became apparent during the development of regulatory guidance for the maintenance rule. The Nuclear Management and Resources Council (NUMARC), now the Nuclear Energy Institute (NEI), representing the industry, asked that the inspection procedure (which is normally developed after the regulatory guidance has been issued) be prepared early and given to the industry for it's use while preparing the industry guidance document. In particular, the industry wanted to use the inspection procedure to address details not in the rule itself.

To develop implementation guidance, the NRC and NUMARC established parallel steering and working groups. In June, 1993, the NRC published RG 1.160, "Monitoring the Effectiveness of Maintenance at Nuclear Power Plants," which endorsed NUMARC 93-01, "Industry Guideline for Monitoring the Effectiveness of Maintenance at Nuclear Power Plants." The staff also developed the draft inspection procedure and validated its use during nine pilot site visits from September, 1994 through March, 1995. Although not specifically required by the rule, NUMARC 93-01 guidance uses PRA importance measures to identify safety (risk) significant SSCs within the scope of the maintenance rule. The guidance specifies that licensees monitor the reliability, availability, and/or condition of all safety (risk) significant SSCs under the scope of the rule. This guidance prescribes both risk-informed and performance-based elements as being acceptable methods to implement the MR.

B.4   Description of the Maintenance Rule

Paragraph (a) of the rule contains most of its detailed technical requirements; paragraph (b) defines the

---

[1]Although Appendix J is a good example of a process-oriented rule, the NRC is revising this rule to add a results-oriented option, thereby permitting licensees the flexibility to adjust leak rate test frequencies on the basis of performance.

[2] Although the maintenance rule has been described as a risk-informed, results-oriented rule, it prescribes certain specific program actions. For example, paragraph (a)(3) requires licensees to perform a periodic evaluation every refueling outage, not to exceed 24 months. In addition, paragraph (b) describes certain safety-related and non-safety-related SSCs that are required to be within the scope of the maintenance rule. Therefore, although the maintenance rule is much more results-oriented and less programmatic than most other existing rules, it has certain process-oriented approaches.

scope of SSCs within the rule; and paragraph (c) states that licensees shall implement the rule no later than July 10, 1996. Paragraph (a) consists of sections (1), (2), and (3).

B.4.1   Goals and Monitoring

Paragraph (a)(1) of the rule requires the operator of each power reactor to set goals and monitor the performance or condition of SSCs in a manner sufficient to give reasonable assurance that those SSCs are capable of performing their intended functions. The rule states that goals must be commensurate with safety and, where practical, shall take into account industry-wide operating experience. The rule also requires licensees to take appropriate corrective action when the performance or condition of an SSC does not meet established goals. In keeping with the non-prescriptive intent of the rule, the licensee, not the NRC, establishes the goals.

B.4.2   Effective Preventive Maintenance

Paragraph (a)(2) of the rule establishes an alternative approach to the monitoring regime required by paragraph (a)(1) of the rule. In this approach, NRC recognizes that, in certain cases, the performance or condition of SSCs could be effectively controlled by doing adequate preventive maintenance rather than by monitoring against goals.

B.4.3   Periodic Evaluations

Paragraph (a)(3) of the rule requires licensees to evaluate the performance and condition-monitoring activities and associated goals and preventive maintenance activities at least every refueling cycle, provided the interval between evaluations does not exceed 24 months. This paragraph requires licensees to systematically review activities under paragraphs (a)(1) and (a)(2) of the rule and to adjust those activities where needed. These evaluations are required by the rule to take IOE into account.

Paragraph (a)(3) of the rule also requires licensees to make adjustments, where necessary, to ensure that the objective of preventing failures of SSCs through maintenance is appropriately balanced against the objective of minimizing the time SSCs are unavailable because of monitoring or preventive maintenance. This requirement recognizes that performing monitoring or preventive maintenance often requires that the SSCs be taken out of service, rendering them unavailable for operation. The higher reliability gained by increased monitoring or preventive maintenance could decrease availability and possibly impair safety.

B.4.4   Safety Assessments Before Performing Maintenance

At the time of the visits to the pilot sites, paragraph (a)(3) of the rule stated that in performing monitoring and preventive maintenance activities, licensees *should* consider an assessment of the total plant equipment that is out of service to determine the overall effect on performance of safety functions. To address this element of the rule, licensees should continually evaluate whether voluntary removal of equipment from service to perform monitoring and preventive maintenance activities may place the plant in a less safe condition, especially if other supportive equipment is out of service. An example of this type of situation might be taking one train of a safety system out of service while one of the alternate sources of power for the redundant train is also out of service. Although TS requirements partially address this concern, the NRC staff identified vulnerabilities that were not addressed by the TSs. The safety assessments should preclude plans for scheduling maintenance in high safety (risk) significant configurations with several pieces of equipment being taken out of service at the same time, even if currently allowed by TSs.

B.4.5   Scope

Paragraph (b) of the rule defines those SSCs that must be included within the scope of the rule. They include all safety-related SSCs and those non-safety-related SSCs that are relied upon to mitigate accidents or transients or are used in EOPs; and whose failure could prevent safety-related SSCs from fulfilling their intended functions; or could cause a scram or safety system actuation.

B.5   Development of Implementation Guidance

To develop implementation guidance, the NRC and NUMARC established parallel steering and working groups. In June 1993, the NRC published RG 1.160, "Monitoring the Effectiveness of Maintenance at Nuclear Power Plants," which endorsed NUMARC 93-01, "Industry Guideline for Monitoring the Effectiveness of Maintenance at Nuclear Power Plants," dated May 1993. NUMARC sponsored two industry workshops in August 1993 to educate the industry on the methods stated in NUMARC 93-01 for implementing the rule. The NRC staff participated in these workshops.

The NRC staff developed a draft inspection procedure to verify implementation of the rule. On March 31, 1994, the NRC sponsored a public workshop in Rockville, Maryland, at which members of the public and the nuclear industry could ask questions about the inspection procedure. At the workshop, NRC explained its expectations about implementation of the rule.

From September 1994 to March 1995, the NRC staff visited nine pilot sites to validate the draft inspection procedure. NRC coordinated with NEI in selecting the following sites: Grand Gulf, Maine Yankee, Shearon Harris, Pilgrim, Byron, Hatch, Vogtle, South Texas, and Crystal River. The licensees for these plants voluntarily implemented most of the requirements of the rule, which did not become effective until July 10, 1996. The NRC review teams included representatives from HQMB and the Probabilistic Safety Assessment Branch from the Office of Nuclear Reactor Regulation (NRR), the Trends and Patterns Analysis Branch from the former Office for the Analysis and Evaluation of Operational Data (AEOD), and regional inspectors.

B.6   Lessons Learned From Visits to Pilot Sites

As documented in NUREG-1526, licensees were generally thorough in determining which SSCs were within the scope of the rule at each site. The visits indicated that use of an expert panel was an appropriate and practical method of determining which SSCs are risk significant. When setting goals, all licensees considered safety, but many did not consider operating experience throughout the industry. Contrary to the requirements of the rule, licensees were not effectively monitoring the performance or condition of certain low safety (risk) significant standby systems at the system or train level. Additionally, most licensees had not established adequate monitoring of structures under the rule. Licensees established reasonable plans for doing periodic evaluations, balancing unavailability and reliability, and assessing the effect of taking equipment out of service for maintenance. However, these plans were not evaluated because they had not been fully implemented at the time of the site visits.

Licensees and inspectors can refer to NUREG-1526 to obtain additional historical information on lessons learned from the pilot site visits. The staff used these lessons learned to document the adequacy of NUMARC 93-01 as endorsed by RG 1.160, and drafted a maintenance rule inspection procedure. The staff also held a public workshop with the industry in June 1995 to discuss lessons learned from the pilot site visits and to give the industry an opportunity to comment on the draft inspection procedure. The NRC also discussed additional industry work needed to complete acceptable MR programs and procedures. From the pilot site visits, the NRC determined that NUMARC 93-01 guidance presented one acceptable method to implement the maintenance rule; however, other acceptable methods to implement the rule were found. After the workshop, NRC completed IP 62706, "The Maintenance Rule," dated August 31, 1995, and began training regional inspectors planning to participate in MRBIs.

APPENDIX C

MAINTENANCE RULE BASELINE INSPECTION PROCESS

C.1   Objectives

The MRBIs were completed between July 15, 1996, and July 10, 1998. The purpose of the MRBIs was to verify licensees' efforts to implement the requirements of the MR in accordance with the guidance in NUMARC 93-01, "Industry Guideline for Monitoring the Effectiveness of Maintenance at Nuclear Power Plants," as endorsed by RG 1.160. The objective of the MRBIs was to verify that licensees established effective maintenance rule programs that complied with the requirements of the rule using Inspection Procedure (IP) 62706, "The Maintenance Rule." Since the MR is the NRC's first risk-informed, performance-based rule, inspectors were instructed to perform program-based inspections on the adequacy of licensees' implementation of the rule instead of the performance-based or results-oriented approach to inspections. This was necessary since the NRC and industry needed to gain experience with risk-informed, performance-based regulations; therefore, the NRC evaluated the adequacy of MR programs, procedures, and implementation practices needed to comply with the prescriptive elements as well as the performance-based risk-informed elements of the rule. The inspectors verified the adequacy of MR programs, procedures, and implementation practices using the inspection guidance provided in IP 62706.

C.2   Need for the MR Baseline Inspections

NRC management envisioned a need to complete MRBIs at every site since this was the first risk-informed, performance-based rule. Risk-informed, results-oriented rulemaking is also being developed in other regulatory areas (e.g., in-service testing, in-service inspection, fire protection, and graded quality assurance). As a result, NRC senior managers determined that MRBIs should be performed at all nuclear power plants so that the NRC and industry would gain experience by implementing this type of regulation. The lessons learned would then be applied to other risk-informed, performance-based regulations as noted above. The NRC also could not rely solely on performance-based inspections until the NRC staff developed confidence that the programs licensees developed to comply with the rule would accurately monitor maintenance performance. In addition, licensees would need to adjust their maintenance activities and programs where performance indicated that a change was necessary.

C.2.1   Pre-Baseline Process

The Commission, senior NRC staff management, and the industry has been concerned that the maintenance rule needs to be inspected and enforced in a consistent manner in all four regions. The Commission stated this concern during a briefing on June 25, 1996, on the status of operating reactors.

The staff took action to ensure consistent inspection and enforcement of the MR. Extensive training was given to the appropriate inspectors and technical staff (especially those inspectors and staff who participated in the baseline inspection program) to give them a uniform understanding of the MR requirements, staff regulatory positions, use of inspection procedures, and inspection planning. Another element of consistency, was the participation of HQMB staff in each baseline inspection. HQMB maintains program responsibility for NRC staff oversight on implementation of the rule.

The staff also established a panel to review MR related findings with potential notices of violations before the inspection report was issued. The panel had members from the Office of Nuclear Reactor Regulation, the Office of Enforcement, and the regions. The panel was described in EGM 96-001, dated July 3, 1996, from James Lieberman, Director, Office of Enforcement, to the regional administrators.

On July 5, 1996, the director of NRR sent a memorandum to remind the regional administrators of the importance for consistent inspection and enforcement of the MR and the need for continued coordination with HQMB. The memorandum required the regional administrators to identify the Senior Executive Service (SES) manager responsible for MR activities in the manager's region. In this capacity, the SES managers attended the first few MRBI exit meetings in their region and other such meetings periodically to ensure that (1) all inspectors completed the required training before conducting a MRBI, (2) the enforcement panel reviewed proposed notices of violation related to the MR, and (3) all communications with licensees on the number of SSCs being monitored in accordance with paragraph (a)(1) of the rule are not inappropriately construed to be a maintenance performance indicator.

C.2.2   Baseline Inspection Schedule

The first MRBI was conducted at Palo Verde Nuclear Power Plant between July 15 and 19, 1996. The team comprised a team leader from HQMB, four senior inspectors (one from each region) performing both horizontal and vertical slice reviews of MR requirements, and several HQMB members serving in the staff support member (SSM) role to maintain team inspection and enforcement consistency. This inspection was viewed as experience for all team members. Following this inspection, the team performed two inspections a month in each region for 3 months. The team also included additional specialist inspectors with PRA expertise in a training mode and two to three SSMs from HQMB. By December, 1996, the NRC regions had completed four inspections a month. A core of MRBI staff was thoroughly trained to conduct MRBIs, and these inspections were being completed at regular intervals. HQMB provided a SSM for each inspection and this continued until all MRBIs were complete by July 10, 1998.

C.2.3   Maintenance Rule Program Reviews

To complete MR program reviews, the staff determined that both horizontal and vertical slice program reviews of licensees' MR procedures and implementation practices were needed to ensure adequate inspector review.

The definition of horizontal slice program reviews essentially means that the inspector verifies that the licensee implemented a particular MR requirement for all SSCs within the scope of the rule. For example, in accordance with paragraph (a)(1) of the rule, licensees must establish goals, monitor SSC performance or condition against established goals, consider industry operating experience when establishing goals, and take corrective actions for SSCs not meeting established goals. To confirm that the licensee implemented this horizontal slice program area, the inspectors verified that the licensee completed these activities for all SSCs within the MR scope that should be monitored under paragraph (a)(1).

Licensees' implementation processes were evaluated for the following horizontal slice program areas: scope of SSCs within the rule, establishing performance measures and goals, determining high and low safety (risk) significant SSCs using PRA, use of the expert panel to determine high and low safety (risk) significant SSCs, and use of IOE. In addition, the extent to which licensees identified FFs, MPFFs, and repetitive MPFFs was evaluated. Finally, inspectors verified that licensees conducted periodic evaluations to balance reliability and availability and performed safety assessments before removing equipment from service for maintenance.

The inspectors performed both horizontal slice program reviews and vertical slice sampling reviews of SSCs within the scope of the rule. The definition of a horizontal slice program review essentially means that the inspector verified that the licensee implemented a particular MR requirement for all SSCs within the scope of the rule. The definition of vertical slice sampling of SSCs within the scope of the rule essentially means that the inspector verified that the licensee implemented all MR requirements for a sample of SSCs within the scope of the rule.

The licensee must implement all MR requirements and the inspection team should complete all inspection requirements listed in IP 62706 for all SSCs within the scope of the rule. Licensees are required to implement all MR requirements for each SSC within the scope of the rule. The inspectors used program implementation guidance in IP 62706 to verify that each MR requirement was met. The inspectors followed this guidance when verifying that all MR requirements were completed for each SSC within the scope of the rule.

C.3   Overall Results From Baseline Inspections

The staff conducted MRBIs at all sites, using IP 62706. The MRBIs were primarily performed by the regions, with a HQMB SSM accompanying each team. The NRC staff completed all MRBIs within the allotted 2-year time, as directed by NRC senior managers. Generally, the NRC found that licensees did a good job of implementing the MR; however, licensees' efforts to implement the MR along with NRC inspection efforts were resource intensive and required a larger number of staff-hours to implement and inspect when compared to other prescriptive or process-oriented regulations. The NRC concluded that the lessons learned and results of risk-informed, performance-based regulations should be incorporated into other risk-informed performance-based regulations, where applicable. The NRC also determined that once the MRBIs were completed, future MR inspections should concentrate on the risk-informed, performance-based elements of the rule in accordance with the new NRC inspection and oversight program.

APPENDIX D

EXAMPLES OF SSCs THAT MAY BE WITHIN THE SCOPE OF THE MAINTENANCE RULE FOR PLANTS IN A DECOMMISSIONING STATUS

- spent fuel pool cladding

- spent fuel pool cooling and cleanup (SFPCC)

- spent fuel pool structure and any connecting piping system seals

- spent fuel pool neutron absorber plates for BWRs (e.g., boral or boraflex)

- spent fuel pool building

- radiation monitors above or in the spent fuel pool

- standby service water (i.e., the portion that is heat sink for SFPCC heat exchangers)

- residual heat removal (i.e., in BWRs) used to support spent fuel pool cooling

- leak detection (This system detects leakage from the spent fuel pool, the transfer pool, and reactor well pool liners. Portions of this system that detect leakage from the spent fuel pool could be within the scope of the maintenance rule.)

- heating, ventilation, and air conditioning system above the spent fuel pool

- standby gas treatment system in BWRs

- spent fuel pool water level instrumentation and control system

- spent fuel pool emergency or normal makeup water supply

- fire water makeup supply used in emergencies

- spent fuel pool crane and equipment (These may need to be included as a result of accident scenarios involving dropped fuel bundles, which could cause cladding damage and potential radiation exposure to personnel in the spent fuel pool area.)

- equipment to support the functions of maintaining required coolant chemistry conditions and reactivity control for PWRs

- standby auxiliary ac power system (i.e., the power supply to SFPCC pumps)

- spent fuel pool transfer tube penetration seals or bellows, pneumatic air system which inflates the seals, and transfer tube gate seals (Failure of these seals could cause a spent fuel pool drain down to a few feet above top of active fuel, the area around the spent fuel pool would become a very high radiation area, and plant staff would need to evacuate the spent fuel pool area.)

APPENDIX E

FUTURE NRC MAINTENANCE RULE ACTIVITIES AND THE DEVELOPMENT OF INSPECTION PROCEDURES TO EVALUATE MAINTENANCE EFFECTIVENESS

IQMB is developing risk-informed performance-based approaches to evaluate licensees' maintenance effectiveness while observing maintenance for plants implementing the monitoring requirements of the maintenance rule (MR). This inspection procedure will instruct resident inspectors to perform vertical slice reviews of plant performance issues for SSCs within the scope of the rule. Such reviews will ensure that licensees maintain living MR monitoring programs, which will achieve NRC's objective of the rule to improve overall plant performance.

The procedure will incorporate risk-informed, performance-based approaches to the routine inspection program for maintenance observations during which the resident inspector staff will focus its attention on HSS and some LSS SSCs experiencing performance problems (i.e., high SSC unavailability due to maintenance, repetitive MPFFs, etc.). The priority will be to focus attention on the safety significance of SSCs experiencing performance problems.

Current Status of the Inspection Procedures for Maintenance Effectiveness

IQMB is revising the inspection procedures (IPs) on maintenance effectiveness with the following objectives:

- Verify that maintenance activities for SSCs within the scope of the MR, including certain non-safety-related SSCs, are being conducted in a manner sufficient to ensure reliable, safe operation of the plant and its equipment and to meet the MR requirements and other regulatory requirements.

- Use a risk-informed, performance-based approach to the inspection of maintenance activities by using plant or SSC performance data (e.g., monitor equipment reliability, unavailability, and emergent maintenance) to prioritize inspection activities.

- Ensure that maintenance-related plant events, reactor trips, and safety system actuations are identified, root cause or cause determination analyses are completed, and corrective actions are taken to prevent recurrence of failures caused by maintenance-related issues.

The results of the risk-informed, performance-based inspection approach may reveal maintenance program implementation issues. If additional inspection of programmatic issues is warranted, inspectors may use regional initiative inspection procedures to further investigate the cause of a performance problem.

NRC FORM 335
(2-89)
NRCM 1102,
3201, 3202

U.S. NUCLEAR REGULATORY COMMISSION

BIBLIOGRAPHIC DATA SHEET

(See instructions on the reverse)

1. REPORT NUMBER
(Assigned by NRC, Add Vol., Supp., Rev., and Addendum Numbers, If any.)

NUREG-1648

2. TITLE AND SUBTITLE

Lessons Learned From Maintenance Rule Baseline Inspections

3. DATE REPORT PUBLISHED

MONTH	YEAR
October	1999

4. FIN OR GRANT NUMBER

5. AUTHOR(S)

S. M. Wong, F. X. Talbot, R. M. Latta, R. P. Correia, T. R. Quay

6. TYPE OF REPORT

Final

7. PERIOD COVERED *(Inclusive Dates)*

7/96 through 7/98

8. PERFORMING ORGANIZATION - NAME AND ADDRESS *(If NRC, provide Division, Office or Region, U.S. Nuclear Regulatory Commission, and mailing address; if contractor, provide name and mailing address.)*

Division of Inspection Program Management
Office of Nuclear Reactor Regulation
U.S. Nuclear Regulatory Commission
Washington, DC 20555-0001

9. SPONSORING ORGANIZATION - NAME AND ADDRESS *(If NRC, type "Same as above"; if contractor, provide NRC Division, Office or Region, U.S. Nuclear Regulatory Commission, and mailing address.)*

Same as Above

10. SUPPLEMENTARY NOTES

11. ABSTRACT *(200 words or less)*

This report summarizes the lessons learned from 68 maintenance rule (MR) baseli ne inspections (MRBIs)conducted at plants with operating licenses in accordance with Title 10, Part 50, of the Code of Fe deral Regulations (CFR) and 3 MRBIs conducted at plants with decommissioning certifications in accordance 10 CFR 50.82(a)(2) . The MRBIs were conducted between July 15, 1996, and July 10, 1998. In general, these MRBIs revealed that licensees i mplemented the requirements of the maintenance rule , 10 CFR 50.65, by following the guidance in NUMARC 93-01, "I ndustry Guideline for Monitoring the Effectiveness of Maintenance at Nuclear Power Plants", as endorsed by NRC Regul atory Guide 1.160, "Montoring the Effectiveness of Maintenance at Nuclear Power Plants." Additionally, licensee s effectively determined which structures, systems, and components (SSCs) at each site were within the scope of the MR. T he use of the expert panels was effective in determining which SSCs were risk significant. The results of the MRBIs also in dicated that the several expert panels performed other MR activities that exceeded the guidance in NUMARC 93-01. Whe n setting goals or performance measures (criteria) in accordance with 10 CFR 50.65(a)(1) or (a)(2), respectively, most licensees considered the risk insights from the probabilisitic risk assessments (PRAs). However, early MRBIs revealed that som e licensees did not have adequate technical justification for deviating from SSC reliability and availability assumptions i n the PRAs when establishing goals or performance criteria and did not adequately assess planned and emergent maintenance activit ies. Most licensees' individual self-assessments were a MR program implementation strength. Most licensees als o established reasonable plans and methods to periodically evaluate the effectiveness of equipment performance and preventive maintenance, including the balance between reliability and availability.

12. KEY WORDS/DESCRIPTORS *(List words or phrases that will assist researchers in locating the report.)*

Maintenance Rule
Maintenance Rule Baseline Inspections

13. AVAILABILITY STATEMENT

unlimited

14. SECURITY CLASSIFICATION

(This Page)

unclassified

(This Report)

unclassified

15. NUMBER OF PAGES

16. PRICE

NRC FORM 335 (2-89)

This form was electronically produced by Elite Federal Forms, Inc.